MAT 025

DEVELOPMENTAL MATHEMATICS

GUIDED NOTEBOOK
THIRD EDITION

for **Fayetteville Technical Community College**

Lead Editor:
Barbara Miller

Editors:
Allison Conger,
Ian Craig,
Jolie Even,
S. Rebecca Johnson

Creative Services Manager:
Trudy Tronco

Designers:
Lizbeth Mendoza,
Patrick Thompson,
Joel Travis

Design and Layout Assistance:
U. Nagesh,
E. Jeevan Kumar,
D. Kanthi,
K.V.S. Anil

A division of Quant Systems, Inc.

546 Long Point Road
Mount Pleasant, SC 29464

Copyright © 2025, 2024, 2020, 2019 by Hawkes Learning / Quant Systems, Inc. All rights reserved.

No part of this publication may be reproduced, stored in a retrieval system, or transmitted in any form or by any means, electronic, mechanical, photocopying, recording, or otherwise, without the prior written consent of the publisher.

Printed in the United States of America 🇺🇸

10 9 8 7 6 5 4 3 2 1

ISBN: 978-1-64277-875-5

Table of Contents

Preface
How to Use the Guided Notebook vi
Strategies for Academic Success ix

CHAPTER 1
Whole Numbers
1.1 Introduction to Whole Numbers. 3
1.2 Addition and Subtraction with Whole Numbers 7
1.3 Multiplication with Whole Numbers 11
1.4 Division with Whole Numbers 15
1.5 Rounding and Estimating with Whole Numbers. 19
1.6 Problem Solving with Whole Numbers 23
1.7 Exponents and Order of Operations 27
1.8 Tests for Divisibility 31
1.9 Prime Numbers and Prime Factorizations 35
Chapter 1 Projects
   Aspiring to New Heights! 39
   Just Between You and Me 41

CHAPTER 2
Fractions and Mixed Numbers
2.1 Introduction to Fractions and Mixed Numbers 45
2.2 Multiplication with Fractions. 51
2.3 Division with Fractions 55
2.4 Multiplication and Division with Mixed Numbers 59
2.5 Least Common Multiple (LCM). 63
2.6 Addition and Subtraction with Fractions 67
2.7 Addition and Subtraction with Mixed Numbers 71
2.8 Comparisons and Order of Operations with Fractions 75
Chapter 2 Projects
   On a Budget 79
   What's Cookin', Good Lookin'? 81

CHAPTER 3
Decimal Numbers
3.1 Introduction to Decimal Numbers 85
3.2 Addition and Subtraction with Decimal Numbers. 89
3.3 Multiplication with Decimal Numbers 93
3.4 Division with Decimal Numbers. 97
3.5 Estimating and Order of Operations with Decimal Numbers 101
3.6 Decimal Numbers and Fractions 105
Chapter 3 Projects
   What Would You Weigh on the Moon? 109
   A Trip to The Grocery Store. 111

CHAPTER 4
Ratios, Proportions, and Percents
- **4.1** Ratios and Unit Rates 115
- **4.2** Proportions 119
- **4.3** Decimals and Percents 123
- **4.4** Fractions and Percents 127
- **4.5** Solving Percent Problems Using Proportions 131
- **4.6** Solving Percent Problems Using Equations 135
- **4.7** Applications of Percent 139
- **4.8** Simple and Compound Interest 143
- **Chapter 4 Projects**
  - Take Me Out to the Ball Game! 147
  - How Much Will This Cell Phone Cost? 149

CHAPTER 5
Measurement
- **5.1** US Measurements 153
- **5.2** The Metric System: Length and Area 155
- **5.3** The Metric System: Capacity and Weight 159
- **5.4** US and Metric Equivalents 163
- **Chapter 5 Projects**
  - Metric Cooking 167
  - Confused Conversions 169

CHAPTER 6
Geometry
- **6.1** Angles and Triangles 173
- **6.2** Perimeter 183
- **6.3** Area 187
- **6.4** Circles 191
- **6.5** Volume and Surface Area 195
- **6.6** Similar and Congruent Triangles 199
- **6.7** Square Roots and the Pythagorean Theorem 205
- **Chapter 6 Projects**
  - Before and After 209
  - This Mixtape Is Fire! 211

CHAPTER 7
Statistics, Graphs, and Probability
- **7.1** Statistics: Mean, Median, Mode, and Range 215
- **7.2** Reading Graphs 219
- **7.3** Constructing Graphs from Databases 225
- **7.4** Probability 229
- **Chapter 7 Projects**
  - What's My Average? 233
  - Misleading Graphs 235

CHAPTER 8
Introduction to Algebra
- 8.1 The Real Number Line and Absolute Value 239
- 8.2 Addition with Real Numbers 243
- 8.3 Subtraction with Real Numbers 247
- 8.4 Multiplication and Division with Real Numbers 251
- 8.5 Order of Operations with Real Numbers 255
- 8.6 Properties of Real Numbers 259
- 8.7 Simplifying and Evaluating Algebraic Expressions 263
- 8.8 Translating English Phrases and Algebraic Expressions 267

Chapter 8 Projects
- Going to Extremes! 271
- Ordering Operations 273

CHAPTER 9
Solving Linear Equations and Inequalities
- 9.1 Solving Linear Equations: $x + b = c$ and $ax = c$ 277
- 9.2 Solving Linear Equations: $ax + b = c$ 281
- 9.3 Solving Linear Equations: $ax + b = cx + d$ 285
- 9.4 Working with Formulas 289
- 9.5 Applications: Number Problems and Consecutive Integers 293
- 9.6 Applications: Distance-Rate-Time, Interest, Average, and Cost 297
- 9.7 Solving Linear Inequalities in One Variable 301
- 9.8 Compound Inequalities 307
- 9.9 Absolute Value Equations 311
- 9.10 Absolute Value Inequalities 315

Chapter 9 Projects
- A Linear Vacation 319
- Breaking Even 321

CHAPTER 10
Graphing Linear Equations and Inequalities
- 10.1 The Cartesian Coordinate System 325
- 10.2 Graphing Linear Equations in Two Variables 331
- 10.3 Slope-Intercept Form 335
- 10.4 Point-Slope Form 341
- 10.5 Introduction to Functions and Function Notation 345
- 10.6 Graphing Linear Inequalities in Two Variables 349

Chapter 10 Projects
- What's Your Car Worth? 353
- Demand and It Shall Be Supplied 355

Math@Work 357
Answer Key 389
Formula Pages 399

How to Use the Guided Notebook

There are a variety of elements in this Guided Notebook that will help you on your way to mastering each topic. Here is a rundown of how to use the elements as you work through this notebook.

Fill-in-the-Blanks

1. When there is an incomplete sentence, you will need to write in the _____*missing*_____ word(s).

   The _____*missing*_____ words can be found by reading through the Learn screens.

Boxed Content

Definitions and **procedures** are highlighted within a box, like the ones shown here. The missing content will vary from box to box. Sometimes an entire definition is missing and sometimes only part of a sentence is missing. Here are two examples of the box variations.

Definition

First term to define: _____*Write the definition here.*_____

Second term to define: _____*If there is another term, define it the same way as above.*_____

DEFINITION

Terms Related to Probability

_____*Outcome*_____	An individual result of an experiment.
_____*Sample Space*_____	The set of all possible outcomes of an experiment.
_____*Event*_____	Some (or all) of the outcomes from the sample space.

DEFINITION

Properties and **Procedure** boxes are completed in a similar way:

> ### Commutative Property of Multiplication
> The order of the numbers in multiplication can be _reversed without changing the product._
> For example, _3 · 4 = 12 and 4 · 3 = 12._
>
> **PROPERTIES**

> ### Subtracting Whole Numbers
> 1. Write the numbers _vertically_ so that the _place values are lined up in columns._
> 2. Subtract only the _digits with the same place value._
> 3. Check by _adding the difference to the subtrahend._ The sum must be _the minuend._
>
> **PROCEDURE**

▶ Watch and Work

For each Watch and Work, you will need to watch the corresponding video in Learn mode and follow along while completing the example in the space provided.

Example 5 Multiplying Whole Numbers

Multiply: 12 · 35

Solution

The standard form of multiplication is used here to find the product 12 · 35.

$$\begin{array}{r} \overset{1}{1}2 \\ \times\ 35 \\ \hline 60 \\ 360 \\ \hline 420 \end{array}$$

12 · 5 = 60
12 · 30 = 360
Product

✏ Now You Try It!

After working along with the example video, work through a similar exercise on your own in the space provided.

Example A Multiplying Whole Numbers

Multiply: 25
          × 42
          1050

1.1 Exercises

Each section has exercises to offer additional practice problems to help reinforce topics that have been covered. The exercises include Concept Check, Practice, Applications, and Writing & Thinking questions. The odd answers can be found in the Answer Key at the back of the book.

Concept Check

True/False. Determine whether each statement is true or false. If a statement is false, explain how it can be changed so the statement will be true. (**Note:** There may be more than one acceptable change.)

1. When the given statement is true, you write "True" for the answer.

   True

Practice

For each set of data, find **a.** the mean, **b.** the median, **c.** the mode (if any), and **d.** the range.

2. The ages of the first five US presidents of the 20th century on the date of their inaugurations were as follows. (The presidents were Roosevelt, Taft, Wilson, Harding, and Coolidge.)

   42, 51, 56, 55, 51

   a. 51   b. 51   c. 51   d. 14

Applications

Solve.

3. Suppose that you have taken four exams and have one more chemistry exam to take. Each exam has a maximum of 100 points and you must average between 75 and 82 points to receive a passing grade of C. If you have scores of 85, 60, 73, and 76 on the first four exams, what is the minimum score you can make on the fifth exam and receive a grade of C?

   81

Writing & Thinking

4. State how to determine the median of a set of data.

   The first step to finding the median is always to arrange the data in order. Once the data is in order, the median is the number in the middle. If there is an even number of items, average the two middle numbers to find the median.

Strategies for Academic Success

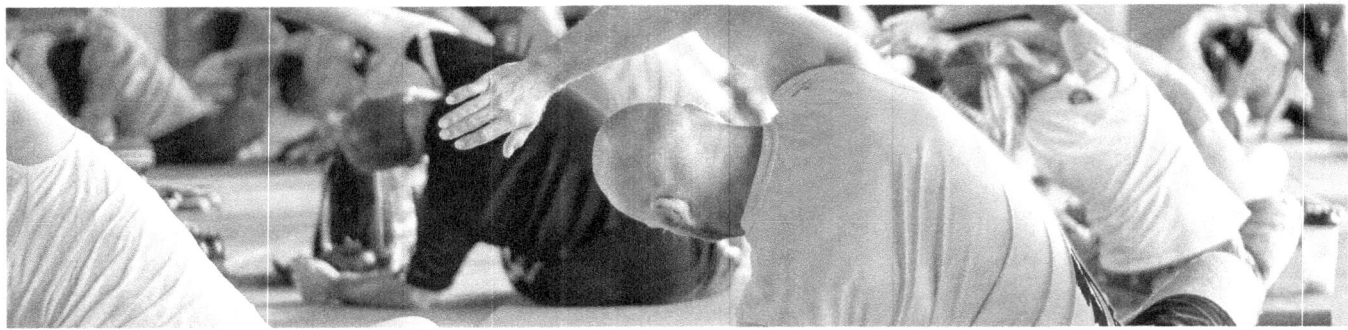

Strategies for Academic Success 🎓

0.1 Understanding and Reducing Stress

Stress is a part of daily life and comes from many different sources, including school, work, and family life. Stress in small doses can be positive, providing you with more energy and making you feel more alert. However, too much stress can start affecting you mentally and physically. You may feel overwhelmed or physically sick as the result of too much stress.

🚩 Determine the Causes of Stress

Stress can be caused by either external or internal stressors. External stressors are things outside of ourselves and are typically situations that we cannot control, such as class registration deadlines, traffic jams, or the loss of a loved one. Internal stressors happen inside of our minds and are related to our feelings, expectations, and goals, such as feeling nervous about an upcoming math exam or annoyed with a friend's decision. Internal stressors can also be the result of decisions we make. While external and internal stressors are separate, they often do affect one another. For instance, if you are feeling nervous about a presentation and run into technology issues, the technology issues can increase your nervousness, which can influence how you react to the situation.

Examples of common stressors are as follows. Consider whether each is an external stressor, internal stressor, or a combination of both.

School: disappointment about a test result, class schedule, homework

Work: hours, commute, finances, budgets, nervousness about a presentation

Family life: childcare, spending time with family, family conflict

🔍 Putting Stress into Perspective

The first step in dealing with stress is to take a step back and evaluate the situation. Try to understand what is causing your stress, and then determine which parts of the situation you can control and which parts you cannot control. Typically, we have control over internal stressors but do not have control over external stressors.

Suppose you have a math final next week that you need to study for. You discover during a meeting with your advisor that you need at least a B on the exam to maintain your scholarship. Besides finding time to study, you also have classes to attend, a job to work, and an important family event, along with all of the regular day-to-day activities in your life, such as sleeping, bathing, and eating. When viewed all together, this situation can seem stressful.

Taking a step back and carefully considering each item listed, we can assign a priority to each and put the whole situation into perspective. To assign a priority, you can use a scale of 1 to 3, in which 1 is the most important, and 3 is the least important. For instance, you might consider studying for the exam and working to be priority 1, attending classes and the family event priority 2, and some of your day-to-day activities (such as catching up on news or podcasts) to be priority 3.

Once you put your stressors into perspective, you can focus on managing your stress.

📅 Plan to Manage Stress Effectively

Without a plan to manage the various stressors in life, it's easy to become overwhelmed. To prevent the stressors from affecting your mental and physical health, it's best to have a plan in place to effectively manage stress.

Take Action: Once the causes of stress are identified and given a priority level, you can take action to manage them. Focus on potential solutions instead of focusing on the problem. Consider creating a prioritized to-do list or creating a weekly schedule. It can also be helpful to determine if certain items are completely out of your control, and if so, letting them go.

Stay Organized: Staying organized can greatly reduce stress. This includes keeping work or school papers organized as well as keeping upcoming events and deadlines organized in a calendar.

Ask for Help: Asking for help can be a stressor itself but can also greatly reduce the stress that a situation or event is causing. The help can come in many forms; you can ask family or friends for assistance with childcare while you study for an exam. You can ask a therapist for help coping with the loss of a loved one. There are a variety of services available to help you deal with nearly any stressor in your life, regardless of whether the stressor is internal or external.

Focus on Yourself: Your physical and mental health play a role in how you respond to any stressor that arises. If you're tired or hungry, your reaction to stressors may be magnified. Getting enough rest, eating healthy meals, and exercising regularly can help. You can also find ways to manage your immediate reactions to stress by practicing breathing techniques, mediating, or doing yoga. With a clear mind and a healthy, well-rested body, you'll have an easier time managing any stressor that comes your way.

Practice Prevention: As Benjamin Franklin once said, "An ounce of prevention is worth a pound of cure." Some stressors can be prepared for in advance. For example, you can review your class notes daily and finish all homework assignments so that you don't need to cram last minute for the final exam. Similarly, you can save money each paycheck for financial emergencies instead of worrying about how to borrow money when an emergency strikes. Sometimes, prevention is simply avoiding procrastination; other times, it requires foresight of potential stressors and preventative action.

Actively Reduce Stress and Anxiety

Practicing deep breathing is an effective strategy to reduce in-the-moment stress or anxiety. During moments of high stress, several things happen in the body: blood pressure and heart rate increase, breathing becomes more shallow and rapid, and the flight-or-fight response can be triggered. By practicing deep breathing, also known as diaphragmatic breathing, you can combat these natural responses. This works because deep breathing triggers the rest and relaxation response, which reduces anxiety, lowers blood pressure, and decreases heart rate.

To practice deep breathing yourself, try the following simple technique:

1. Breathe in slowly, until your lungs feel full, while imagining drawing the air towards your belly.
2. Hold in this breath for a few seconds.
3. Slowly exhale all the air out of your lungs.
4. Repeat for several cycles.

If this method of deep breathing does not work well for you, try another. Any deep-breathing technique can trigger the same relaxation response. Try practicing deep breathing if you are experiencing high anxiety before or during a math test or during any other stressful events in your day-to-day life.

> **Questions**
> 1. List two internal stressors and two external stressors in your life right now. Give each stressor a priority level.
> 2. Create a plan to deal with the stressors you listed in Question 1.

Strategies for Academic Success 🎓

0.2 Staying Organized

Staying organized is a key element in reducing stress and making sure you finish all your tasks. This is true whether you are a student, employee, parent, or in any other stage of life. It is especially true if you are in multiple stages at once.

While keeping a mental to-do list is an option, it's likely that you'll run into a variety of issues with this method, such as forgetting to do tasks, scheduling overlapping appointments, or missing a loan payment. Occasional issues like these might not seem like a big deal, but if they are recurring, you may eventually end up failing a class, losing your job, or paying late fees.

📘 Keep a Planner

A planner is a place for you to organize your schedule and record any important tasks or responsibilities. By keeping all of this information in one place, you'll be less likely to forget to complete something important.

You can make a planner or you can purchase one. Planners can be made in either printed form or electronic form. Planners can be broken down into daily, weekly, or monthly pages. Daily planners allow for the most detail per day, while monthly planners only allow for a high-level overview of events. Weekly planners are a balance between the two.

When purchasing or creating a planner, consider how much detail you'll need in order to keep your life organized. If you need hourly breakdowns of each day, a daily planner will be best. If you need less precision, but still have several events or tasks each day, then a weekly planner might suit your needs. If you only need to keep track of a few events per week or month, then a monthly planner might be the option that works for you.

It is worth noting that an electronic planner can travel with you everywhere you carry your phone. There are a variety of apps available for most operating systems. At the very least, the calendar app on your phone can be utilized to remind you of the most important events in your schedule.

If you are unsure which type of planner you need, list out the regular events and tasks you have coming up, and then determine the level of precision you'll need to accomplish everything on your list.

Items on your list might include the following:
- Your work schedule
- Your class schedule
- Academic calendar dates
- Assignment deadlines
- Billing due dates
- Important holidays and birthdays
- Upcoming trips

When filling out your planner, first prioritize the events and tasks, then schedule the most important items before adding any nonessential items. Once you've filled in your planner, be sure to keep it up to date by adjusting appointments or adding in new events. And don't forget to always check your planner before scheduling any new appointments!

🗂 Use a File System

A well-organized file system is important for keeping documents sorted and easy to find at any moment. File systems can be used for a variety of documents, such as course materials, financial papers, and legal documents.

Keeping track of course materials, such as handouts, homework, exams, and notes, can help you improve your grades. If these materials are readily available, it is easier to fill free time with reviewing and studying. Other important papers to keep track of are course policies and the course syllabus. Having these documents on hand will help you know what to expect in the course along with a list of important dates.

The first step in creating a file system is determining what is important to save. If you save everything, your file system may become cluttered and unusable. If you keep too few items, your file system may not be informative enough to be helpful. A good rule of thumb is to keep recent important documents and discard older documents after a certain date. For instance, you may decide to keep all current course documents (notes, homework, tests, and so on) while the course is in progress, then trim down the files to only the notes and tests once the course is complete. For financial documents, you may decide to keep initial loan or account information, recent account statements, and account updates, then decide to discard old files after a certain number of years.

The second step in keeping an organized file system is deciding where to save files. Printed documents can be saved in binders, folders, or divided cabinets. For example, you can have a binder that holds all of your course information, with a divider tab for class information, notes, homework, and tests. Keeping the files within each section in order by date can make it easy to find what you are looking for when you need it.

Electronic files can be saved in a system of nested folders. For example, you can have a main folder titled with the course name, and within that folder, have individual subfolders for notes, tests, and course documents. It's important to clearly name files that are saved electronically.

✂ Create a Workspace

An organized workspace can help you focus and work efficiently. If you have a dedicated workspace that is comfortable and uncluttered, it is easier to sit down and work without distractions. Having enough space is also essential. For example, if you are working on math homework, you need space for your computer, textbook, notebook, calculator, and writing utensils. If you have a clean area where you can spread out, you are less likely to misplace items when you need them.

If your workspace needs to be used for multiple purposes, having an easy-to-access storage space for your work items is essential. Keeping your items organized and together can help you set up and break down your workspace as needed.

If you live in a crowded house or dorm room, finding a workspace at home can be tricky. You may need to get a little creative and be flexible. The items you need at your workspace should ideally be easily relocated from room to room and completely stored away when not in use. This will allow the workspace to be multifunctional.

If you don't have space to work where you live, you can consider public places such as a library or a coffee shop. Consider the pros and cons of each location before making a choice. Libraries are typically free to use, but you might be limited on hours you can be in the building or how long you can use study rooms. Coffee shops typically require you to make a purchase to stay for any length of time, but the music and atmosphere can help create a focused mindset. Both options usually have free Wi-Fi, but proceed with caution when using unsecured public Wi-Fi and don't access important personal information when connected.

It is also important to figure out the level of distractions you can tolerate before picking a location to set up your workspace. Distractions can come in the form of friends and family talking to you, an open space on campus getting too crowded, or the smell of baked goods at a bakery becoming too tempting. While it's unlikely you can avoid all distractions, keeping them to a minimum can help make your study session more productive.

Questions

1. Would you benefit best from a daily, weekly, or monthly planner? Explain your reasoning.

2. Describe your current workspace and determine three changes you can make to create a more organized workspace.

Strategies for Academic Success 🎓

0.3 Managing Your Time Effectively

Have you ever made it to the end of a day and wondered where all your time went? Sometimes it feels like there aren't enough hours in the day. Finding time to balance work, school, and home life can be difficult. Some interruptions, like unexpected traffic or family emergencies, are simply outside of your control. However, other distractions are within your control, such as watching TV or scrolling through social media. It's important to find a balance between activities you need to do (such as attending class and work) and activities you want to do (such as watching TV). Managing your time is important because you can never get time back.

Here are three strategies for managing your time more effectively.

⚖️ Take Breaks

When you are working on an important project or studying for a big exam, you may feel tempted to work as long as possible without taking a break. This is especially true when you're working or studying at the last minute. While staying focused is important, working yourself for hours until you're mentally drained will lower the quality of your work and force you to take even more time recovering.

Think about the way that overworking can affect your body physically. If you're weight-training, you must take frequent breaks both between individual sets and entire workout sessions. If you don't let your muscles recover, you risk injuring yourself, which could leave you laid up for weeks.

Just like taking breaks helps your physical body recover, it will also help your brain re-energize and refocus. During study sessions, you should plan to take a short study break at least once an hour. If you usually work indoors, take this time to get a breath of fresh air outside and clear your head.

Study breaks and work breaks should usually last around five minutes. The longer the break, the harder it is to start working again. Instead of stopping for half an hour, take a five-minute break and reward yourself with some downtime when the task is complete. Similarly, if a course you are taking has a built-in break during the middle of the class period, use it to get up and move around. This little bit of physical movement can help you think more clearly.

📰 Avoid Multitasking

Multitasking is working on more than one task at a time. When you have several assignments that need to be completed, you may be tempted to save time by working on two or three of them at once. While this strategy might seem like a time-saver, you will probably end up using more time than if you had completed each task individually. Not only will you have to switch your focus from one task to the next, but you will also make more mistakes that will need to be corrected later.

People don't multitask nearly as well as they think they do. For example, research studies have shown that multitasking while driving is similar to or even worse than driving while drunk. While multitasking on a project for school or work may not be dangerous, it can lead to wasted time and silly mistakes. Instead of trying to do two things at once, schedule yourself time to work on one task at a time.

Multitasking can also become an excuse for distractions, especially electronic ones. Have you ever tried to complete a homework assignment, watch TV, and message friends all at the same time? You probably did one of these things well and two of these things badly. That's because your brain can't give its full attention to three tasks at once. To stay focused in class or while studying, try stashing your phone in your backpack or purse and staying logged out of your computer until you need it.

⏲ Use a Time Budget

Just like a financial budget shows you how you spend your money, a time budget shows you how you spend your time. A time budget can help you identify "wasted" time that could be used more productively.

To begin budgeting your time, you first need to get an idea of how you usually spend it. For one week, keep track of everything you do in fifteen-minute time blocks. Be as accurate as possible and track as you go throughout your day. Here's an example of what a partial record of activities for one day might look like.

Time	Activity
12:00 a.m. to 7:30 a.m.	Sleeping
7:30 a.m. to 8:15 a.m.	Getting ready for class
8:15 a.m. to 8:45 a.m.	Driving
8:45 a.m. to 9:00 a.m.	Walking to class
9:00 a.m. to 10:30 a.m.	Math class
10:30 a.m. to 10:45 a.m.	Walking to next class
10:45 a.m. to 11:30 a.m.	English class

Once you've recorded your entire week in fifteen-minute time blocks, you can calculate how much time you spend on different types of activities. First, review the activities you entered and assign each of them to one of the following categories: sleep, meals, work, class, study, extracurricular, exercise, personal, family, entertainment, social, and other. Then, answer the following questions.

1. Do you feel like you had enough time to fulfill all your responsibilities?
2. Were you surprised by how much time you spent on any particular activity?
3. What are some important activities you should have spent more time doing?
4. What are some activities you'd like to spend more time doing in the future?
5. What are some activities you'd like to spend less time doing in the future?

Based on your answers to these questions, create a weekly time budget. Remember that one week contains only 168 hours. If you want to spend more time on a particular activity, you'll need to find that time somewhere. Use a planner to schedule specific blocks of time for study sessions, meals, travel times, and morning/evening routines. As a general rule, you should set aside at least two hours of study time for every one hour of class time. That means that a three-credit-hour course would require at least six hours of outside study time per week.

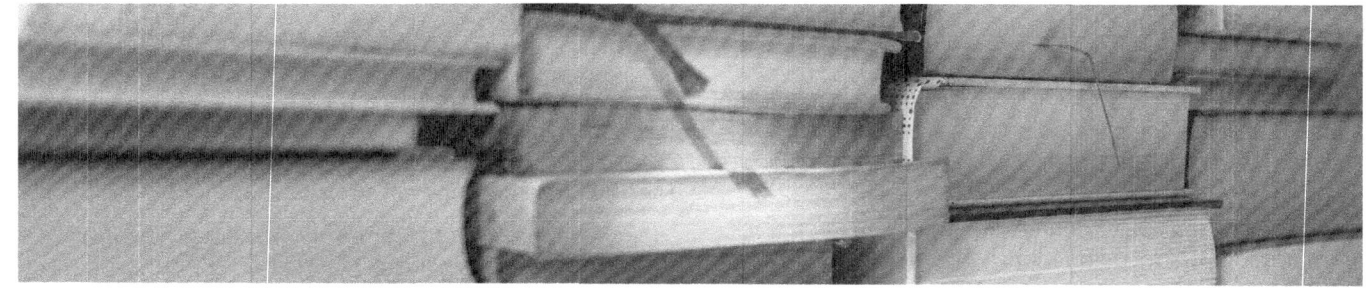

Strategies for Academic Success 🎓

0.4 Reading a Textbook and Note-Taking

Reading a math textbook and taking notes is more intensive than casually reading a book for fun. You have to concentrate more on what you are reading and be careful how you take notes because you will likely be tested on the content. Here are some tips to help you successfully read a math textbook and take notes.

📖 Reading a Math Textbook

Reading a math textbook requires a different approach than reading literature or history textbooks because math textbooks contain a lot of symbols and formulas in addition to words. Whether you are reading a physical math textbook or an e-book, the following tips can help you understand and retain the information presented.

Start at the Beginning: Don't start in the middle of an assigned section. Math tends to build on previously learned concepts, and you may miss an important concept or formula that is crucial to understanding the rest of the material in the section.

Don't Skim: When reading math textbooks, look at everything: titles, learning objectives, definitions, formulas, text in the margins, and any text that is highlighted, outlined, or in bold. Also, pay close attention to any tables, figures, charts, and graphs.

Work through Examples: Make sure you understand each step of an example. If you don't understand something, mark it so you can ask about it in class. Sometimes, math textbooks leave out intermediate steps to save space. Try working through the examples on your own, filling in any missing steps.

Understand the Mathematical Definitions: Many terms used in everyday English have a different meaning when used in mathematics. Some examples include *equivalent*, *similar*, *average*, *median*, and *product*. It is important to note these differences in your notebook along with other important definitions and formulas. You might find it helpful to keep a separate math glossary that contains each term with its definition along with an example of how the term is used.

Reread important information: After reading through the section, go back and reread boxed definitions, examples, or any other important information. It is also important to reread any portions of the lesson that seemed more complex or difficult compared to the rest.

✏️ Annotating a Textbook

Annotating involves marking the text and taking notes in the margin. Put your book to good use and don't be afraid to add comments and highlighting. If you don't understand something in the text, reread it a few times. If it is still not clear, note the text with a question mark or some other notation so that you can ask your instructor about it. A well-annotated text can help you find important information while completing homework assignments as well as help you review for exams.

You can use sticky notes, pens, and highlighters to annotate a physical text. Try adding special colors or symbols for different types of information. For example, questions could be in blue with a question mark, and important ideas could be in red with a star. However, if you are renting a textbook, be sure to limit annotations to sticky notes and easily removed markings. Most e-books can be annotated with mark-up tools available in the e-reader.

Some important things to annotate or highlight are definitions, key concepts, important ideas, and examples. You can also add notes about prior knowledge to the margin and summarize definitions or procedures in your own words.

✔ General Note-Taking Tips

- Write the date and the course name at the top of each page.
- Write the notes in your own words and paraphrase.
- Use abbreviations, such as ft for foot, # for number, def for definition, and RHS for right-hand side.
- Copy all figures or examples that are presented during the lecture.
- Review your notes after class and add additional detail where needed.
- Review and rewrite your notes after class. Do this on the same day, if possible.

There are many different methods of note taking and it's always good to explore new methods. A good time to try out new note-taking methods is when you rewrite your class notes. Try each new method a few times before deciding which works best for you. Presented here are three note-taking methods you can try out. You may even find that a blend of several methods works best for you.

📄 Note-Taking Methods

Taking notes in class is an important step toward understanding new material. Review your notes daily while you do your homework and before taking quizzes and tests. While there are several methods for taking notes, three are presented here.

Outlines

An outline consists of several topic headings, each followed by a series of indented bullet points that include subtopics, definitions, examples, and other details.

Example:
1. Ratio
   a. Comparison of two quantities by division.
   b. Ratio of a to b
      i. $\frac{a}{b}$
      ii. $a:b$
      iii. a to b
   c. Can be reduced
   d. Common units can cancel

Split Page

The split page method divides the page vertically into two columns, with the left column narrower than the right column. Main topics go in the left column, and detailed comments go in the right column. The bottom of the page is reserved for a short summary of the material covered.

Example:

Keywords:	Notes:
Ratios	1. Comparison of two quantities by division 2. $\frac{a}{b}$, $a:b$, a to b 3. Can reduce 4. Common units can cancel

Summary: Ratios are used to compare quantities and units can cancel.

Mapping

The mapping method is the most visual of the three methods. One common way to create a mapping is to write the main idea or topic in the center and draw lines from that main idea to smaller ideas or subtopics. Additional branches can be created from the subtopics until all of the key ideas and definitions are included. Using a different color for subtopics can help you visually organize the topics.

Example:

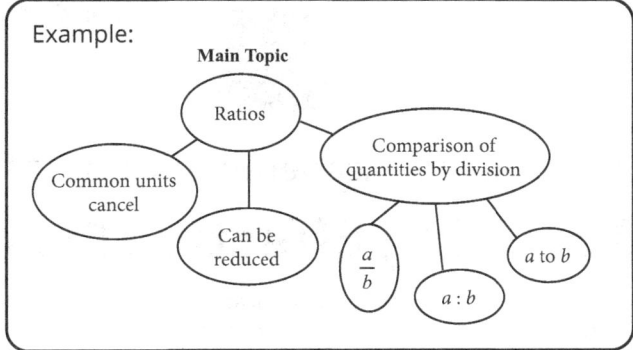

📖 Reviewing Supplemental Content

Many textbooks have corresponding online courseware that includes lesson or example videos to supplement the text. The videos can vary in topic from reviewing the skills and concepts covered in the lesson to working through a specific example. While watching these videos, keep the following tips in mind.

- Pause the video as needed to self-assess your understanding.
- Avoid other audio or visual distractions.
- Take notes.
- Write down any additional questions asked during the video, then try to find the solution in the textbook or ask your professor for more information.
- Rewatch videos as needed for clarification.
- Work through the same or similar exercises on paper while watching the video or soon after.

In addition to videos, you might have access to other supplemental content such as PowerPoints, chapter reviews, and content created by your instructor.

Questions

1. Explain why you need to carefully read a math textbook in the order the information is presented.
2. Which note-taking method do you use? (If it isn't one listed, describe it.)

Strategies for Academic Success

0.5 Using Effective Study Strategies

Have you ever heard the phrase "practice makes perfect"? This saying applies to many things in life. You won't become a concert pianist without many hours of practice. You won't become an NBA basketball star by sitting around watching basketball on TV. You can watch all of the videos and read all of the books on how to do something, but you won't learn the skills without actually practicing. The same idea applies to math: math is not a spectator sport.

Just as you work your body through physical exercise, you have to work your brain through mental exercise. Math is an excellent subject to provide the mental exercise needed to stimulate your brain. So when doing mathematics, remember the 3 P's—Practice, Patience, and Persistence—and the positive effects they will have on your brain! A few ways to instill practice, patience, and persistence into your routine are to manage your study schedule, use creative study strategies, and diligently prepare for exams.

Manage Your Study Schedule

Creating a study routine will keep you in the habit of studying and provide personal accountability. Studying daily will also help prevent the need for emergency cram sessions before an exam. To make studying part of your daily routine, try using the following strategies.

Find a study time that works for you. If you're a morning person, wake up early for a brief study session. If you struggle to stay awake after eating lunch, don't plan to study in the early afternoon.

Schedule study times in your planner. Adding study sessions to your planner makes them harder to skip. Plan to study two to three hours outside of class for every hour spent in class. Schedule your most difficult subjects first so your mind is fresh.

Be flexible when necessary. Unexpected events are bound to happen once in a while. Don't feel guilty about rescheduling your study time around these disruptions. You should also be flexible with your study location and not be afraid to relocate if you start getting distracted by your surroundings.

Keep study times separate. When you have multiple assignments due for class, you may find yourself spending all your study time working on assignments and not reviewing. Even if you only have fifteen minutes to spare, reserve time for studying your class notes.

Review class notes as soon as possible. Reviewing class notes soon after class will help you retain the skills and concepts covered. This will also allow you to fill in any gaps in your notes for future reference.

Use Creative Study Strategies

Sometimes studying can be boring and you lose your attention after extended periods of time. Creative study strategies can help retain your attention and help you learn.

Recite information aloud. Ask yourself questions about the material to see if you can recall important facts and details. Pretend you are teaching or explaining the material to someone else.

Use mnemonics or memory techniques. For example, a mnemonic that is commonly used to remember the order of operations is "Please Excuse My Dear Aunt Sally," which uses the first letter of the words Parentheses, Exponents, Multiplication, Division, Addition, and Subtraction.

Use acronyms to help remember important concepts or procedures. An acronym is created by taking the first letter (or letters) from each word in the phrase that you want to remember and making a new word. For example, the acronym HOMES is often used to remember the five Great Lakes in North America, where each letter in the word represents the first letter of each lake: Huron, Ontario, Michigan, Erie, and Superior.

Use visual images like diagrams, charts, and pictures. You can make your own pictures and diagrams to help you recall important definitions, theorems, or concepts.

Split larger pieces of information into smaller "chunks." For example, instead of remembering a sequence of digits such as 555777213, you can break it into chunks and remember 555 777 213.

Group long lists of information into categories that make sense. For example, instead of remembering all the properties of real numbers individually, try grouping them into shorter lists by operation, such as addition and multiplication.

Associate the information with something you already know. Think about how you can make the new information personally meaningful. How does it relate to your life, your experiences, and your current knowledge? If you can link new information to existing memories, you can create "mental hooks" to help you recall the information in the future.

Tips for Preparing for an Exam

When preparing for a final math exam, it's best to take at least a week to prepare for the exam along with regular studying throughout the semester.

Determine Important Exam Information

1. What is the date, time, and location of the exam?
2. Is there a time limit to complete the exam?
3. What materials can you bring to the exam? Can you use formula sheets, calculators, or scrap paper?

One Week before the Exam

1. **Create a study schedule.** Determine where to study and whether you will study with classmates. Be sure to limit distractions when deciding. Bring snacks and take regular breaks.
2. **Organize your study materials.** Even if you have well-organized course materials, you may decide to condense them into a single study sheet, create note cards, or make a formula sheet.
3. **Follow through with your study plan.** Starting right away will give you time to ask your instructor or classmates for help with difficult topics.

Three Days before the Exam

1. Make a practice test and take it under the same constraints the final exam will have.
2. Ask your instructor or classmates about any questions you struggled with on the practice test.

Night before the Exam

1. Organize and pack all of the supplies you will need for the exam: pencils, erasers, calculator, scratch paper, and so on.
2. Review your formula sheet.
3. Avoid trying to cram in a lot of last-minute studying.
4. Go to bed early and get a good night of sleep.

Day of the Exam

1. Eat a healthy breakfast and don't drink too much caffeine, which can make you anxious.
2. Make sure you have all of your supplies with you when you leave for the exam.
3. Review your formula sheet if you are not allowed to use it on the exam.
4. Get to the exam location early so you can be organized and mentally prepared.

Questions

1. Which creative study strategy do you find most useful, and why? (If it isn't one listed, describe it.)
2. Describe your current method for preparing for an exam and determine three ways you can improve it.

Strategies for Academic Success 🎓

0.6 Reducing Test Anxiety

Approximately 93% of adults in the United States have experienced anxiety related to math at some point in their life.[1] Anxiety can come in many forms when it comes to the subject of math. You may have general anxiety about math, you may have anxiety about specific aspects of math class, or you may have test anxiety. Practicing the following strategies can help you reduce any general math anxiety and test-specific anxiety you may experience.

🏆 Reducing Math Anxiety

Depending on what causes your math anxiety, the tools to overcome it will differ. You may need to develop an effective study strategy so that you always feel prepared for your math class or any test. You may need to learn how to properly read a math textbook and take notes. You may even need to learn to manage your time, prioritizing what is most important (such as work and studying), or develop a system to keep your documents and workspace organized.

Math anxiety could also be caused by your thoughts and feelings about either your abilities or the subject itself (or both). If this is the case, try the following methods to reduce math anxiety.

1. Maintain a positive attitude and avoid negative self-talk. Remind yourself that each mistake is an opportunity to learn and improve. Set small math achievement goals to keep you moving toward bigger goals.
2. Visualize yourself doing well in math, whether it's on a quiz, test, or passing the course.
3. Think of setbacks as opportunities for growth. When you experience a setback, learn from your mistakes or the situation and keep moving forward.
4. Take deep, slow breaths, and think about the people or places that make you feel happy and peaceful.
5. Find a ritual that makes you happy. This could be listening to music, taking deep breaths, or imagining your future success. Find what works for you and practice it regularly.
6. Form a math study group. Working with others will help you feel more relaxed, and you can support each other.
7. Learn effective study skills that work with your learning style. Don't be afraid to change the way you study if it's not working for you.

📋 Preparing for an Exam

Being prepared for an exam can reduce your anxiety leading up to and during the exam. Preparing for an exam is not just the studying that happens the few days before the exam. The preparation should take place throughout the semester. Be sure to consistently do the following to always be prepared for exams or pop quizzes.

1. Read your math textbook before class and review the lesson again after class.
2. Take notes using the method that works best for you, and review the notes regularly. Rewrite your notes if necessary to expand or condense the information covered.
3. Keep your class materials organized. Use labels or highlighting so you can easily review the content.
4. Use study aids, such as note cards, to help you remember definitions, theorems, formulas, or procedures.
5. Actively practice the skills you learn by working through the examples and doing your homework.
6. Strive to understand the material, not simply memorize it. Explain the material in your own words or look for patterns.

---

[1] Christie Blazer, "Strategies for Reducing Math Anxiety. Information Capsule. Volume 1102," ERIC, September 2011, eric.ed.gov/?id=ED536509.

7. Plan to study two to three hours outside of class for every hour spent in class. Pick a study time and place that works well with your schedule.

8. Do not spend all of your study time working through a single exercise or figuring out a single concept. Move on to the next exercise or concept. Make notes to ask your instructor or classmates for help when needed.

» Test-Taking Strategies to Reduce Anxiety

In addition to preparing for a test by studying, you can try the following tips to reduce your test anxiety.

1. Get plenty of sleep the night before the exam and eat nutritious meals on the day of the exam. Being well-rested and avoiding hunger will help you focus.

2. Talk to your instructor about your anxiety. Your school might allow special accommodations, such as extra time on the test or taking the test in a more calming area.

3. When handed a test, immediately write your name at the top, and then perform a "brain drain" to write down all the formulas and important facts you remember on your test or scratch paper. Having this information readily available will boost your confidence and reduce your anxiety.

4. Carefully read the directions of the exam before starting any work. Then, read through the questions to make sure you understand them. You can add notation next to each question to indicate the perceived level of difficulty. It might build your confidence to start with problems you find easier and then move on to the more difficult problems.

5. If you panic or freeze during a math test, focus on a single problem you can do. Once you gain confidence, work through other problems you know how to do. Then, attempt the more difficult problems.

6. Check your solutions as you go to make sure they make sense. Then, use any extra time after you finish the exam to do a final check of your work and solutions before you turn it in.

7. If you get to a problem you don't know how to solve, skip it and come back after you finish the problems you do know how to solve. Another problem on the test may help you remember how to solve the more difficult problems.

8. If a problem is multiple choice, work the problem before looking at the answers. Alternatively, you can start by looking at the answer choices and working backwards to see if any are easily eliminated.

🔍 Reviewing Test Results

Whether you ace the test, receive a passing score, or receive a low score, you should take the time to review your work and any feedback from your instructor. This is especially true for tests and exams early in the course because math courses tend to be cumulative. That is, concepts and skills learned for one test are likely to be needed for the next test. When you receive a test back, do the following:

1. Correct any of your incorrect work so that you know the correct way to solve the problem in the future. If you are unsure what you did wrong, ask a classmate or visit your instructor during office hours.

2. Analyze the test questions to determine if most of them came from your class notes, the homework, or the textbook. This will give you an idea of how to spend your time studying for the next test.

3. Analyze the errors you made on the test. Were they careless mistakes? Did you run out of time? Did you not understand the material well enough? Were you unsure which method to use?

4. Based on your analysis of the test, adjust your study methods and schedule. It is important to adjust your study strategy or try new approaches. This may seem difficult if you are used to your current study techniques. Adjusting one or two aspects of your study methods and schedule at a time can allow you to find out what works (or does not work).

Questions

1. Describe any anxiety you may have in relation to math.

2. Create a plan to help work through your math anxiety.

🛟 Support

If you have questions or comments, we can be contacted as follows:

24/7 Chat: chat.hawkeslearning.com

Phone: 1-800-426-9538

E-mail: support@hawkeslearning.com

Web: support.hawkeslearning.com

CHAPTER 1

Whole Numbers

1.1 Introduction to Whole Numbers

1.2 Addition and Subtraction with Whole Numbers

1.3 Multiplication with Whole Numbers

1.4 Division with Whole Numbers

1.5 Rounding and Estimating with Whole Numbers

1.6 Problem Solving with Whole Numbers

1.7 Exponents and Order of Operations

1.8 Tests for Divisibility

1.9 Prime Numbers and Prime Factorizations

CHAPTER 1 PROJECTS
Aspiring to New Heights!
Just Between You and Me

Connections

Numbers play a role in any job you may take in the future. If you go into sales, you will need to keep track of the number of sales you make and the profit you've earned for the company. If you go into construction, you will need to be able to accurately take measurements of the buildings and structures you help create. If you go into the medical field, numerical accuracy is important in situations such as measuring out the medicine for a sick child or recording a patient's blood pressure. If you decide to start your own business, you will need to evaluate operation costs to determine how much to charge your clients for your products or services.

During your education, numbers will also play a large role. For example, as a student striving to succeed, you will want to keep track of your progress in the courses you take. One way to do this is to find the average of your test scores throughout the semester. The average of these scores will help you determine if you are doing well in the course or if you need to seek assistance from either the instructor or a tutor. Suppose during the semester you receive the following scores on the first four tests in a course.

78, 85, 94, 83

What is your current test average for the course? Should you seek assistance to improve your grade?

Name: _____ Date: _____ 3

1.1 Introduction to Whole Numbers

Whole Numbers

The **whole numbers** are _natural numbers_ along with _number 0_

Natural numbers = $\mathbb{N}$ = { _1 2 3 4 5 6 7 8 9 10_ }

Whole numbers = $\mathbb{W}$ = { _0 1 2 3 4 5 6 7 8 9 10_ }

DEFINITION

The Decimal System

The **decimal system** (or base ten system) is a place value system that depends on three things.

1. _The tens digit_
2. the placement of _each digit_
3. the value of _each place_

DEFINITION

1. In **expanded notation**, the values represented by each digit are _____

▶ Watch and Work

Watch the video for Example 3 in the software and follow along in the space provided.

Example 3 Writing Numbers in Expanded Notation

Write each number in expanded notation.

a. 954

b. 6507

Solution

✏️ Now You Try It!

Use the space provided to work out the solution to the next example.

Example A Writing Numbers in Expanded Notation

Write each number in expanded notation.

a. 463

b. 7300

Reading and Writing Whole Numbers

You should note the following four things when reading or writing whole numbers.

1. Digits are read in _____ (groups _____).

2. Commas are used to _____ if a number has _____

3. The word **and** does not _____
   _____

4. Hyphens are used to write words for _____
   _____

1.1 Exercises

Concept Check

True/False. Determine whether each statement is true or false. If a statement is false, explain how it can be changed so the statement will be true. (**Note:** There may be more than one acceptable change.)

1. In the number 21,057, the "1" represents 1000.

2. 56,317 can be written as 56,000 + 300 + 17 in expanded notation.

3. 42,360 can be written as forty-two thousand, three hundred sixty.

4. The word "and" is not used when reading or writing whole numbers.

Practice

5. Name the place value of each nonzero digit in the following number: 24,608.

6. Write 1892 in expanded notation.

7. Write 683,100 in words.

Write each number in standard notation.

8. Write four hundred thousand, seven hundred thirty-six in standard notation.

Applications

9. The largest lake in the United States is Lake Superior. It takes up an area of 82,103 square kilometers. Write 82,103 in words.

10. The largest collection of Joker playing cards consists of eight thousand, five hundred twenty cards amassed by Tony De Santis after inheriting a two thousand piece collection from the magician Fernando Riccardi. Write eight thousand, five hundred twenty in standard form.

Writing & Thinking

11. How are natural numbers and whole numbers different and how are they the same?

12. When are hyphens used to write numbers in English words?

Name: _____ Date: _____   7

1.2 Addition and Subtraction with Whole Numbers

Adding Whole Numbers

1. Write the numbers _____ so that the _____

2. Add only the _____

PROCEDURE

Carrying When Adding Whole Numbers

If the sum of the digits in one column is more than 9,

1. write the _____ and

2. carry the _____ as a number to be added to _____

PROCEDURE

Commutative Property of Addition

The order of the numbers in addition can be _____

For example, _____

PROPERTIES

Associative Property of Addition

The grouping of the numbers in addition can be _____

For example, _____

PROPERTIES

Additive Identity Property

The sum of a number and 0 is _____

For example, _____

The number 0 is called the _____

PROPERTIES

1. The **perimeter** of a geometric figure is _____

8   1.2   Addition and Subtraction with Whole Numbers

▶ Watch and Work

Watch the video for Example 6 in the software and follow along in the space provided.

Example 6 Calculating the Perimeter of a Polygon

Calculate the perimeter of the polygon.

(**Note:** A 5-sided polygon is called a **pentagon**.)

Solution

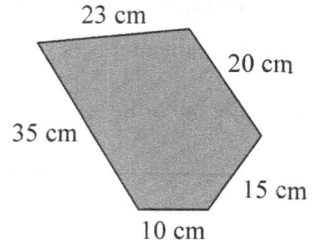

✏ Now You Try It!

Use the space provided to work out the solution to the next example.

Example A Calculating the Perimeter of a Polygon

Calculate the perimeter of the polygon.

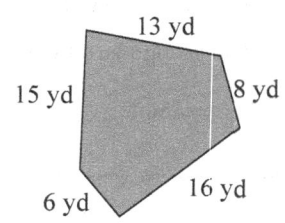

Subtraction

Subtraction is the operation of _____

The **difference** is the result of _____

_____

DEFINITION

Subtracting Whole Numbers

1. Write the numbers _____ so that the _____
2. Subtract only the _____
3. Check by _____ The sum must be _____

PROCEDURE

1.2 Exercises

Concept Check

True/False. Determine whether each statement is true or false. If a statement is false, explain how it can be changed so the statement will be true. (**Note:** There may be more than one acceptable change.)

1. A polygon is a geometric figure in a plane with two or more sides.

2. To find the perimeter of a rectangle, add the lengths of the four sides.

3. When subtracting, sometimes the digit being subtracted is larger than the digit it is being subtracted from and so "carrying" must occur.

4. If your bank account has a balance of $743 and you want to withdraw $115, you would use subtraction to find that the new balance would be $628.

Practice

Simplify.

5. 15
    +43

6. 981
    +46

7. 275
    −131

8. 543
    −167

9. Calculate the perimeter of the given geometric figure.

   9 cm
   4 cm
   5 cm
   12 cm
   7 cm
   5 cm

Applications

Solve.

10. The Magley family has the following monthly budget: $815 mortgage; $69 electric; $47 water; and $122 phone bills (including cell phones). What is the family's budget for each month for these expenses?

11. A couple sold their house for $135,000. They paid the realtor $8100, and other expenses of the sale came to $800. If they owed the bank $87,000 for the mortgage, what were their net proceeds from the sale?

Writing & Thinking

12. Explain when "carrying" should be used in addition with whole numbers and give an example.

13. Explain when "borrowing" would be used in subtraction and give an example.

Name: _____ Date: _____   **11**

1.3 Multiplication with Whole Numbers

Commutative Property of Multiplication

The order of the numbers in multiplication can be _____

For example, _____

PROPERTIES

Associative Property of Multiplication

The grouping of the numbers in multiplication can be _____

For example, _____

PROPERTIES

Multiplicative Identity Property

The product of any number and 1 is _____

For example, _____

The number 1 is called the _____

PROPERTIES

Multiplication Property of 0 (or Zero-Factor Law)

The product of a number and 0 is _____

For example, _____

PROPERTIES

The Distributive Property

Multiplication can be _____

For example, _____

PROPERTIES

1.3 Multiplication with Whole Numbers

▶ Watch and Work

Watch the video for Example 5 in the software and follow along in the space provided.

Example 5 Multiplying Whole Numbers

Multiply: 93 · 46

Solution

✏ Now You Try It!

Use the space provided to work out the solution to the next example.

Example A Multiplying Whole Numbers

Multiply: 15
$\underline{\times\ 32}$

Multiplying Whole Numbers by Powers of 10

To multiply a whole number:

by 10, write _____

by 100, write _____

by 1000, write _____

by 10,000, write _____

and so on.

PROCEDURE

Area of a Rectangle

The **area** of a rectangle (measured in square units) is found by _____

DEFINITION

1.3 Exercises

Concept Check

True/False. Determine whether each statement is true or false. If a statement is false, explain how it can be changed so the statement will be true. (**Note:** There may be more than one acceptable change.)

1. The numbers being multiplied are called the divisors.

2. According to the multiplicative identity, $1 \cdot 25 = 52$.

3. According to the distributive property, $4 \cdot (7+2) = 4 \cdot 7 + 4 \cdot 2$.

4. The associative property of multiplication indicates that length can be multiplied by width or width can be multiplied by length to get the same answer.

Practice

5. Multiply: 42
             $\times 56$

6. Multiply: $40 \cdot 2000$

1.3 Exercises

7. State the property of multiplication illustrated and show that the statement is true by performing the multiplication: $3 \cdot (1 \cdot 7) = (3 \cdot 1) \cdot 7$

8. Rewrite 7(8 + 4) by using the distributive property, then simplify.

9. Calculate the area of the given rectangle.

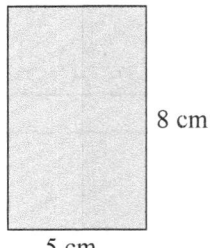

8 cm

5 cm

Applications

Solve.

10. A network television station has approximately 18 minutes of commercial time in each hour. How many minutes of commercial time does the network have in a one-day programming schedule of 20 hours? In one week?

11. A sandwich shop buys 372 loaves of bread for the week. If each loaf of bread has 24 slices, how many slices of bread were purchased?

Writing & Thinking

12. Explain, in your own words, what the zero-factor law indicates.

13. Explain, in your own words, why 1 is called the multiplicative identity.

Name: _____ Date: _____ **15**

1.4 Division with Whole Numbers

1. We can use the division sign (÷) to indicate the division procedure as follows.

$$12 \div 4 = 3$$ Read "12 divided by 4 equals 3."

_____ ÷ _____ = _____

2. Two other notations that indicate division are the following.

_____ → $4\overline{)12}^{\,3}$ ← _____   _____ → $\dfrac{12}{4} = 3$ ← _____

Division by 1
Any number divided by one _____

Example: _____

PROPERTIES

Division of a Number by Itself
Any nonzero number _____

Example: _____

PROPERTIES

Division Involving 0
Case 1: Any nonzero whole number _____

Example: _____

Case 2: _____

Example: _____

PROPERTIES

---

3. The long division process can be written in the following format.

_____

$$6\overline{)27}^{\,4}$$

$$\underline{-24} \leftarrow 6 \cdot 4 = 24$$

$$27 - 24 = 3 \rightarrow 3$$

_____
_____
_____

▶ Watch and Work

Watch the video for Example 3 in the software and follow along in the space provided.

Example 3 Dividing Whole Numbers

Divide: $683 \div 7$

Solution

✏ Now You Try It!

Use the space provided to work out the solution to the next example.

Example A Dividing Whole Numbers

Divide: $415 \div 6$

1.4 Exercises

Concept Check

True/False. Determine whether each statement is true or false. If a statement is false, explain how it can be changed so the statement will be true. (**Note:** There may be more than one acceptable change.)

1. If a division problem has a nonzero remainder, then the divisor and quotient are factors of the dividend.

2. $13 \div 1 = 13$

3. $12 \div 0 = 12$

4. $\dfrac{0}{7}$ is undefined.

Practice

Divide.

5. $13\overline{)0}$

6. $0\overline{)51}$

7. $12\overline{)108}$

8. $11\overline{)4406}$

Applications

Solve.

9. One pint of Ben and Jerry's Crème Brûlée Ice Cream has 64 grams of fat. If there are 4 servings per pint, how many grams of fat are in each serving?

10. US Astronaut Peggy Whitson orbited the Earth 6032 times during her space flights on the International Space Station. If the International Space Station orbits the Earth 16 times per day, how many days was Peggy Whitson in space?

Writing & Thinking

11. Explain how you would check a division problem that has a nonzero remainder.

12. Discuss how division is related to multiplication.

Name: _____ Date: _____   19

1.5 Rounding and Estimating with Whole Numbers

Rounding Numbers

To **round** a given number means _____

DEFINITION

Rounding Rule for Whole Numbers

1. Look at the single digit just to the right of the digit in the place of desired accuracy.

    a. **If this digit is less than 5**, leave the digit in the place of desired accuracy as it is, and _____
    _____

    b. **If this digit is 5 or greater**, increase the digit in the desired place of accuracy by _____
    _____ All digits to the left remain unchanged unless
    _____ Then the 9 is replaced by 0 and _____
    _____.

PROCEDURE

Estimating a Sum or Difference

1. Round each number to _____

2. Perform the _____

PROCEDURE

▶ Watch and Work

Watch the video for Example 4 in the software and follow along in the space provided.

Example 4 Estimating Sums of Whole Numbers

Estimate the sum; then find the actual sum.

$$\begin{array}{r} 68 \\ 925 \\ +487 \\ \hline \end{array}$$

1.5 Rounding and Estimating with Whole Numbers

Solution

✏️ Now You Try It!

Use the space provided to work out the solution to the next example.

Example A Estimating Sums of Whole Numbers

Estimate the sum; then find the actual sum.

$$\begin{array}{r} 176 \\ 84 \\ +\ 75 \\ \hline \end{array}$$

> **Estimating a Product**
> 1. Round each number to _____
> 2. Multiply the _____
>
> PROCEDURE

> **Estimating a Quotient**
> 1. Round both the divisor and dividend to _____
> 2. Divide with _____
>
> PROCEDURE

1.5 Exercises

Concept Check

True/False. Determine whether each statement is true or false. If a statement is false, explain how it can be changed so the statement will be true. (**Note:** There may be more than one acceptable change.)

1. Rounding means finding a number close to the given number, using a specified place of accuracy.

2. When rounded to the ten thousands place, 435,613 becomes 400,000.

3. To estimate the answer for a division problem, begin by rounding both the divisor and dividend.

4. If estimated, 4250 ÷ 51 is 4000 ÷ 50 = 80.

Practice

Estimate each answer; then find the actual answer.

5. 83
   62
   + 78

6. 63,504
   − 42,700

7. 420
   × 104

8. 11)99

1.5 Exercises

Applications

Solve.

9. College expenses for a private four-year college in the 2022–2023 academic year were as follows:

Tuition & Fees	$25,243
Room & Board	$8996
Books & Supplies	$1077

   Estimate the total cost to attend for a year using rounded numbers to the nearest thousand. Then calculate the actual cost.

10. Ramón is running a sand volleyball tournament soon and must purchase some new equipment. He needs three new nets, which cost $159 each. He also needs five new sets of boundary lines, which cost $86 each. Estimate the total cost of the new equipment. Then calculate the actual cost.

Writing & Thinking

11. In your own words, define estimation.

12. Compare and contrast rounding and estimating.

Name: _____ Date: _____ **23**

1.6 Problem Solving with Whole Numbers

> **Basic Strategy for Solving Word Problems**
> 1. READ: _____
> 2. SET UP: Draw any type of figure or diagram that might be helpful and _____ _____
> 3. SOLVE: _____
> 4. CHECK: Check your work and _____
>
> PROCEDURE

> **Finding the Average of a Set of Numbers**
> 1. Find the _____
> 2. Divide this sum by the _____
>
> PROCEDURE

▶ Watch and Work

Watch the video for Example 7 in the software and follow along in the space provided.

Example 7 Calculating an Average

Find the average of the following set of numbers: 15, 8, 90, 35, 27.

Solution

✏ Now You Try It!

Use the space provided to work out the solution to the next example.

Example A Calculating an Average

Find the average of the following set of numbers: 18, 29, 6, 33, 14, 26.

1.6 Exercises

Concept Check

True/False. Determine whether each statement is true or false. If a statement is false, explain how it can be changed so the statement will be true. (**Note:** There may be more than one acceptable change.)

1. Averages are found by performing addition and then division.

2. The sum of 312 and 4 is 1248.

3. The word "quotient" indicates multiplication.

4. After reading a problem carefully, the next step might be to make a diagram or draw a figure.

Applications

Solve.

5. Steven is calculating how many calories are in his lunch. He has a hamburger that has 354 calories, a medium fry that has 365 calories, and a chocolate milk shake that has 384 calories. How many total calories is his meal?

6. For a class in statistics, Anthony bought a new graphing calculator for $95, special graphing paper for $8, a USB flash drive for $10, a textbook for $105, and a workbook for $37. How much did he spend for this class?

7. A square that is 10 inches on a side is placed inside a rectangle that has a width of 20 inches and a length of 24 inches. What is the area of the region inside the rectangle that surrounds the square? (Find the area of the shaded region in the figure.)

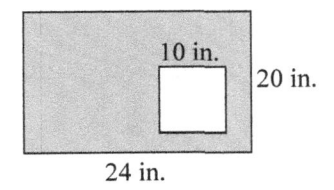

8. The Lee family spent the following amounts for groceries: $338 in June; $307 in July; $318 in August. What was the average amount they spent for groceries in these three months?

Writing & Thinking

9. Make up three word problems that include keywords to indicate operations such as addition, subtraction, multiplication, and division. Underline the keywords.

10. Give an example where you might use average (other than in a class).

1.7 Exponents and Order of Operations

1. When looking at $3^5 = 243$, 3 is the _____, 5 is the _____, and 243 is the _____. _____ are written slightly to the right and above the _____. The expression 3^5 is _____.

The Exponent 1

Any number raised to the first power _____

For example, _____

DEFINITION

The Exponent 0

Any nonzero number raised to the 0 power _____

For example, _____

Note: The expression 0^0 _____

DEFINITION

Rules for Order of Operations

1. Simplify within grouping symbols, such as _____ (If there are more than one pair of grouping symbols, start with _____ _____

2. Evaluate any _____

3. Moving from left to right, perform any _____ _____

4. Moving from left to right, perform any _____ _____

PROCEDURE

---

2. A well-known mnemonic device for remembering the rules for order of operations is the following.

Please	Excuse	My	Dear	Aunt	Sally
↓	↓	↓	↓	↓	↓
_____	_____	_____	_____	_____	_____

▶ Watch and Work

Watch the video for Example 6 in the software and follow along in the space provided.

Example 6 Using the Order of Operations with Whole Numbers

Simplify: $2 \cdot 3^2 + 18 \div 3^2$

Solution

✏ Now You Try It!

Use the space provided to work out the solution to the next example.

Example A Using the Order of Operations with Whole Numbers

Simplify: $6^2 \div 9 + 3 - 14 \div 7$

1.7 Exercises

Concept Check

True/False. Determine whether each statement is true or false. If a statement is false, explain how it can be changed so the statement will be true. (**Note:** There may be more than one acceptable change.)

1. Nine squared is equal to eighteen.

2. $2^7 = 128$

3. 7^0 is undefined.

4. According to the order of operations, multiplication is always performed before division.

Practice

For each exponential expression **a.** identify the base, **b.** identify the exponent, and **c.** evaluate the exponential expression.

5. 2^3

6. 4^0

Simplify.

7. $18 \div 2 - 1 - 3 \cdot 2$

8. $30 \div 2 - 11 + 2(5-1)^3$

Applications

Solve.

9. Neville bought 15 boxes of trading cards. Each box has 10 packs of trading cards. Each pack of trading cards contains 20 cards. He adds 132 cards that he already owns to the newly purchased cards. Then, Neville evenly distributes all of the cards to 6 of his friends. How many trading cards would each person get?

   a. If you simplify the expression $15 \cdot 10 \cdot 20 + 132 \div 6$ using the order of operations, will you get the correct answer? If not, explain what is wrong with the expression.

   b. What is the answer? If necessary, write the corrected expression to get the correct results when following the order of operations.

10. Robert is purchasing shirts for his weekend soccer team. The shirts he wants to buy are normally $25 each but are on sale for $10 off. His team has a total of 11 players. How much will he spend to buy the shirts?

    a. If you simplify the expression $25 - $10 · 11 using the order of operations, will you get the correct answer? If not, explain what is wrong with the expression.

    b. What is the answer? If necessary, write the corrected expression to get the correct results when following the order of operations.

Writing & Thinking

11. Give one example where addition should be completed before multiplication.

Name: _____ Date: _____ 31

1.8 Tests for Divisibility

Divisibility

If a number can be divided by another number so that the remainder is 0, then we say

1. the number is _____
2. the divisor _____

DEFINITION

Divisibility by 2

A number is divisible by 2 (is an **even number**) if _____

DEFINITION

Divisibility by 3

A number is divisible by 3 if _____

DEFINITION

▶ Watch and Work

Watch the video for Example 2 in the software and follow along in the space provided.

Example 2 Determining Divisibility by 3

Determine whether each of the following numbers is divisible by 3.

a. 6801

b. 356

Solution

1.8 Tests for Divisibility

✏ Now You Try It!

Use the space provided to work out the solution to the next example.

Example A Determining Divisibility by 3

Is 7912 divisible by 3? Explain why or why not.

Divisibility by 4

A number is divisible by 4 if _____

DEFINITION

Divisibility by 5

A number is divisible by 5 if _____

DEFINITION

Divisibility by 6

A number is divisible by 6 if _____

DEFINITION

Divisibility by 9

A number is divisible by 9 if _____

DEFINITION

Divisibility by 10

A number is divisible by 10 if _____

DEFINITION

1.8 Exercises

Concept Check

True/False. Determine whether each statement is true or false. If a statement is false, explain how it can be changed so the statement will be true. (**Note:** There may be more than one acceptable change.)

1. A number that is divisible by 10 is also divisible by 2 and 5.

2. 6801 is divisible by 9.

3. 7605 is divisible by 10.

4. 5,187,042 is divisible by 3.

Practice

Using the tests for divisibility, determine which of 2, 3, 4, 5, 6, 9, and 10 (if any) will divide exactly into each given number.

5. 105

6. 150

7. 331

8. 1234

Applications

Solve.

9. You are on a team that is participating in a charity walk with a goal to raise $12,400. Each team member agrees to raise the same amount of money. If the possible team sizes are 5, 6, 9, or 10 members, which team sizes allow the goal amount to be evenly split between the team members? How much money would each team member raise for each team size that can evenly split the goal amount?

10. A company is working on a project that will take 440 hours of work to complete. The manager in charge of the project has the option to have 4, 6, or 8 people work on the project. If the manager wants to evenly divide the work between the team members, which team size will evenly split the work hours? How many hours would each team member spend on the project for each team size that evenly splits the work hours?

Writing & Thinking

11. **a.** If a number is divisible by both 3 and 5, then it will be divisible by 15. Give two examples.

    b. However, a number might be divisible by 3 and not by 5. Give two examples.

    c. Also, a number might be divisible by 5 and not 3. Give two examples.

1.9 Prime Numbers and Prime Factorizations

Prime Number

A **prime number** is a counting number _____

_____

DEFINITION

Composite Number

A **composite number** is _____

DEFINITION

Determining Whether a Number is Prime

Divide the number by progressively larger prime numbers (2, 3, 5, 7, 11, and so forth) until one of the following is true.

1. The remainder _____ This means that the _____
   _____

2. You find a quotient _____ This means that the _____
   _____

PROCEDURE

The Fundamental Theorem of Arithmetic

Every composite number has _____

THEOREM

Finding the Prime Factorization of a Composite Number

1. Factor the composite number _____

2. Factor each _____

3. Continue this process until all factors are prime.

PROCEDURE

1.9 Prime Numbers and Prime Factorizations

> **Factors of a Composite Number**
>
> The only factors (or divisors) of a composite number are
>
> 1. _____
> 2. _____
> 3. products formed by _____
>
> **DEFINITION**

▶ Watch and Work

Watch the video for Example 10 in the software and follow along in the space provided.

Example 10 Finding the Factors of a Composite Number

Find all the factors of 60.

Solution

✏ Now You Try It!

Use the space provided to work out the solution to the next example.

Example A Finding the Factors of a Composite Number

Find all the factors of 42.

1.9 Exercises

Concept Check

True/False. Determine whether each statement is true or false. If a statement is false, explain how it can be changed so the statement will be true. (**Note:** There may be more than one acceptable change.)

1. A prime number has exactly 1 factor.

2. A composite number has 2 or more factors.

3. 231 is a prime number.

4. All the factors of 30 are 1, 2, 3, 5, 6, 10, 15 and 30.

Practice

Determine whether each number is prime or composite. If the number is composite, find at least three factors of the number.

5. 47

6. 63

Find the prime factorization of each number. Use the tests for divisibility for 2, 3, 4, 5, 6, 9, and 10 whenever they help to find beginning factors.

7. 125

8. 150

Applications

Solve.

9. Twenty-four pencils are to be distributed evenly between the members of a group. What are the possible group sizes if each person in the group is to receive the same number of pencils?

10. A chocolatier makes 72 specialty truffles. She wants to sell packages that each have the same number of truffles. What are her options for the number of truffles that can be in a package?

Writing & Thinking

11. Are all odd numbers also prime numbers? Explain your answer.

12. Explain the difference between factors of a number and multiples of that number.

Chapter 1 Project

Aspiring to New Heights!
An activity to demonstrate the use of whole numbers in real life.

You may have never heard of the Willis Tower, but it once was the tallest building in the United States. This structure was originally named the Sears Tower when it was built in 1973, and it held the title of the tallest building in the world for almost 25 years. The name was changed in 2009 when Willis Group Holdings obtained the right to rename the building as part of their lease for a large portion of the office space in the building.

The Willis Tower, which is 1451 feet tall and located in Chicago, Illinois, was the tallest building in the Western Hemisphere until May 2, 2013. On this date a 408 foot spire was placed on the top of One World Trade Center in New York to bring its total height to a patriotic 1776 feet. One World Trade Center now claims the designation of being the tallest building in the United States and the Western Hemisphere.

1. The Willis Tower has an unusual construction. It is comprised of 9 square tubes of equal size, which are really separate buildings, and the tubes extend to different heights. The footprint of the Willis Tower is a 225 foot by 225 foot square. In answering the questions below, be sure to use the correct units of measurement on your answers.

    a. Since the footprint of the Willis Tower is a square measuring 225 feet on each side and it is comprised of 9 square tubes of equal size, what is the side length of each tube? (It might help to draw a diagram.)

    b. What is the perimeter around the footprint, or base of the Willis Tower?

    c. What is the area of the base of the Willis Tower?

    d. What is the area of the base for one square tube?

    e. Write the values from parts c. and d. in words.

2. Suppose there are plans to alter the landscape around Willis Tower. The city engineers have proposed adding a concrete sidewalk 6 feet wide around the base of the building. A drawing of the proposal is shown. (**Note:** This drawing is not to scale.)

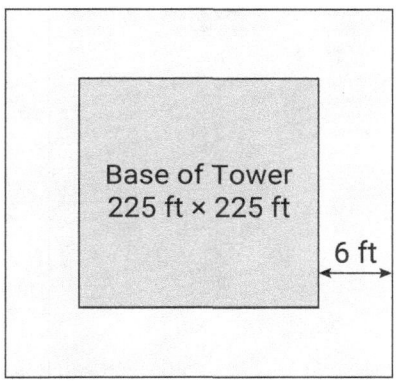

    a. Determine the total area of the base of the tower including the new sidewalk.

    b. Write down the area of just the base of the tower that you determined in part 1c.

    c. Determine the area covered by the concrete sidewalk around the building. (**Hint:** You only want the area between the two squares.)

    d. If a border were to be placed around the outside edge of the concrete sidewalk, how many feet of border would be needed?

    e. If the border is only sold by the yard, how many yards of border will be needed? (**Note:** 1 yard = 3 feet.)

    f. Round the value from part e. to the nearest ten.

    g. Round the value from part e. to the nearest one hundred.

Chapter 1 Project

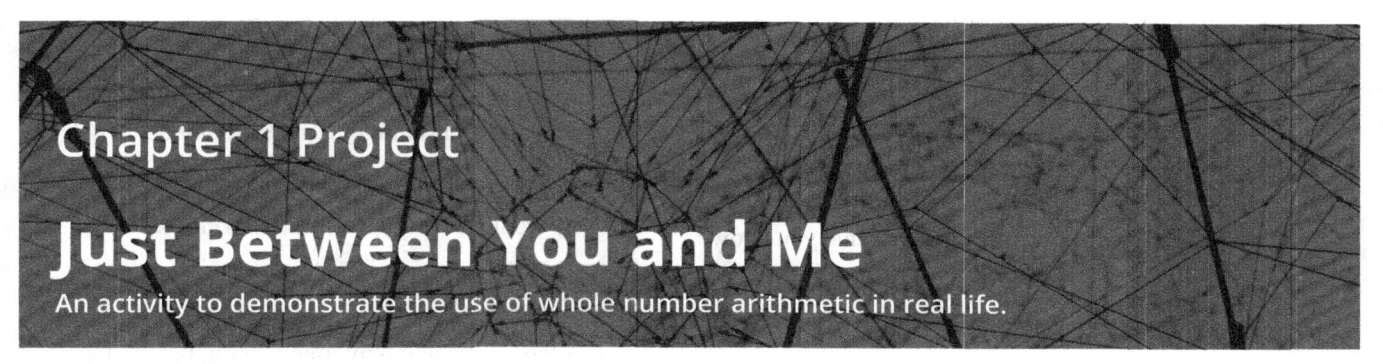

Just Between You and Me
An activity to demonstrate the use of whole number arithmetic in real life.

Cryptography deals with turning a message from plaintext into a code known as ciphertext. Ciphertext appears to the general public as nonsense but can be decoded by the intended recipient. Over the past few hundred years, the field of cryptography has advanced greatly. Around 2000 years ago, all that was usually needed to encode (or encrypt) a message was whole number arithmetic.

To work with this classic form of encryption, often called a shift cipher, there is just one additional thing we need to know. Shift ciphers work similar to how a 12-hour clock shows the same time every 12 hours, even though it's a different time, or even a different day. Extending this concept to the alphabet, Table 1 assigns each letter a number, starting with 0, and repeats the cycle after the letter Z.

A	B	C	D	E	F	G	H	I	J	K	L	M	N	O	P	Q	R	S	T	U	V	W	X	Y	Z	A	B	C	...
0	1	2	3	4	5	6	7	8	9	10	11	12	13	14	15	16	17	18	19	20	21	22	23	24	25	26	27	28	...

Using subtraction, we can find that the numbers representing the two As are 26 apart. Similarly, the two Bs are 26 apart. Just like the digits on a clock begin to repeat after 12 hours, the letters will begin to repeat after 26 letters.

1. Notice C is represented by 28.
   a. Determine the remainder when 28 is divided by 26.
   b. Compare the value of the remainder found in part a. to the values in Table 1. What do you notice about this value and 28?

The idea behind a *shift* cipher is to shift the letters in the plaintext by a given number of letters to form the ciphertext. This encryption can be done with a right-shift or a left-shift.

A *right-shift* is obtained by adding the same value to each of the numbers associated with the letters. Since each letter in Table 1 is associated with a whole number, we will add the value of the right-shift, and then divide by 26, using the letter associated with the *remainder*.

For example, we will apply a right-shift by 14 to encrypt the plaintext HELLO:

Step 1: H, E, L, L, O is turned into 7, 4, 11, 11, 14.

Step 2: Adding 14 to each of these results in 21, 18, 25, 25, 28.

Step 3: When dividing by the 26, the remainders are 21, 18, 25, 25, 2, so the ciphertext is VSZZC.

2. Encrypt the word ZYGOTE using a right-shift by 3.

A *left-shift* is obtained by subtracting the same value from each of the numbers associated with the letters. We will first add 26 to the number associated with the letter, subtract the value of the left-shift, and then divide by 26, using the letter associated with the *remainder*.

For example, to decrypt the ciphertext VSZZC (which was encrypted using a right-shift by 14), we will apply a left-shift by 14:

Step 1: VSZZC becomes 21, 18, 25, 25, 2. Adding 26 to each of these results in 47, 44, 51, 51, 28.

Step 2: Subtracting 14 from each of these results in 33, 30, 37, 37, 14.

Step 3: When dividing by 26, the remainders are 7, 4, 11, 11, 14, so the plaintext is HELLO.

Notice that we added 26 to each number during Step 1. This is a necessary step when applying a left-shift to avoid negative numbers.

3. When the message has more than one word, all letters are capitalized and there are no spaces nor punctuation. This adds to the look of the text being nonsense, but for someone that knows how the plaintext was encoded, part of decoding will be properly separating things out. Suppose a colleague sends the following encrypted message, and you know the colleague used a right-shift by 8.

    QABPMUMMBQVOIBVWWV

    a. Convert the letters in the ciphertext to numbers using Table 1.
    b. What would it take to undo a right-shift by 8? State this and state what numbers are obtained.
    c. Convert these back to letters using Table 1.
    d. Separate out the plain text into individual words to completely recover your colleague's original message.

Suppose you are not the intended recipient of the following ciphertext, but you have intercepted the message. You know that a right-shift by a certain amount has been used, but you do not know by how much.

EWWLEWSLLZWUGXXWWKZGHSLFAFW

For very short messages, a "brute force" approach could be used to check every possible right-shift and undo that shift. But that would mean trying 25 different possibilities. This would be too time-consuming for messages moderately long to very long.

4. Why would there 25 possibilities to check?
5. Instead of checking every possible right-shift, messages can often be decoded using the fact that E is the most commonly used letter in the English language.

    a. Count how many times each letter occurs in the intercepted ciphertext.
    b. Assume that the letter that shows up most frequently in the ciphertext was the letter E in the plaintext. What was the amount of the right-shift?
    c. Based on your answer from part b., decrypt the message. Separate out the plain text into individual words to completely recover the plaintext of the original message you intercepted.

CHAPTER 2

Fractions and Mixed Numbers

2.1 Introduction to Fractions and Mixed Numbers

2.2 Multiplication with Fractions

2.3 Division with Fractions

2.4 Multiplication and Division with Mixed Numbers

2.5 Least Common Multiple (LCM)

2.6 Addition and Subtraction with Fractions

2.7 Addition and Subtraction with Mixed Numbers

2.8 Comparisons and Order of Operations with Fractions

CHAPTER 2 PROJECTS
On a Budget
What's Cookin', Good Lookin'?

Connections

To be a well-informed citizen, it's important to pay attention to politics and the political process since they affect everyone in some way. In local, state, and national elections, registered voters make choices about various ballot measures and who will represent them in the government. In major issues at the state and national levels, pollsters use mathematics (in particular, statistics and statistical methods) to indicate attitudes and to predict how the electorate will vote, typically accurate within certain percentage ranges. When there is an important election in your area, read news articles and listen to news reports for mathematically-related statements predicting the outcomes.

Certain states, such as Ohio, are considered swing states. This means that during presidential elections no single candidate has an overwhelming support from the voters in the state. Swing state status is based on elections in recent history. Ohio is an important swing state because, since the 1960s, the candidate that won the electoral votes in Ohio has typically also won the presidential election.

Suppose that to determine funding for a political advertising campaign, a candidate's campaign staff takes a survey of voters in Ohio. They determine there are 8000 registered voters in a certain precinct, and $\frac{3}{8}$ of the voters are undecided about who they will vote for. The survey indicates that $\frac{4}{5}$ of these undecided voters agree with the candidate's platform. From this information, the campaign staff can determine how many people can be influenced to vote for their candidate with a series of campaign advertisements. How many undecided voters agree with the candidate's platform?

2.1 Introduction to Fractions and Mixed Numbers

1. Numbers such as $\frac{2}{3}$ (read "two-thirds") are said to be in _____.

2. The top number, 2, is called the _____ and the bottom number, 3, is called the _____.

Proper Fractions and Improper Fractions

A **proper fraction** is a fraction in which the _____

Examples: _____

An **improper fraction** is a fraction in which the _____

Examples: _____

DEFINITION

Variable

A **variable** is _____

DEFINITION

The Number 0 in Fractions

For any nonzero value of b,

_____

For any value of a,

_____

DEFINITION

---

3. To graph the fraction $\frac{2}{3}$ proceed as follows.

   1. Divide the interval (distance) from _____
   2. Graph (or shade) the _____

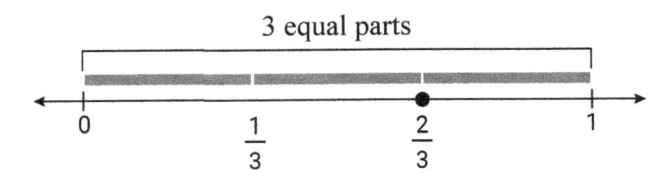

2.1 Introduction to Fractions and Mixed Numbers

4. A **mixed number** is _____

5. To graph the mixed number $1\frac{3}{4}$ proceed as follows.

   1. Mark the intervals from _____

   2. Graph (or shade) the
      _____

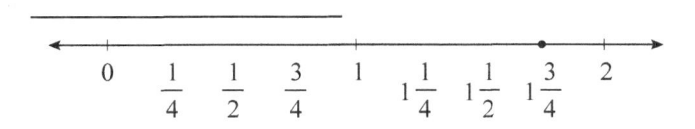

      0   $\frac{1}{4}$ $\frac{1}{2}$ $\frac{3}{4}$ 1 $1\frac{1}{4}$ $1\frac{1}{2}$ $1\frac{3}{4}$ 2

Changing a Mixed Number to an Improper Fraction

1. Multiply the whole number by _____
2. Add the numerator of the _____
3. Write this sum _____

PROCEDURE

▶ Watch and Work

Watch the video for Example 12 in the software and follow along in the space provided.

Example 12 Changing Mixed Numbers to Improper Fractions

Change $8\frac{9}{10}$ to an improper fraction.

Solution

✏️ Now You Try It!

Use the space provided to work out the solution to the next example.

Example A Changing Mixed Numbers to Improper Fractions

Change $10\frac{4}{9}$ to an improper fraction.

> **Changing an Improper Fraction to a Mixed Number**
>
> 1. Divide the numerator by _____ The quotient is _____
> 2. Write the remainder _____
>
> PROCEDURE

2.1 Exercises

Concept Check

True/False. Determine whether each statement is true or false. If a statement is false, explain how it can be changed so the statement will be true. (**Note:** There may be more than one acceptable change.)

1. In $\frac{11}{13}$, the denominator is 11.

2. $\frac{0}{6} = 0$

3. $\frac{17}{0}$ is undefined.

48   2.1   Exercises

Practice

For the figure, **a.** write the fraction for the number of days remaining in June (not crossed out) and **b.** write the fraction for the number of days that have been crossed out for June.

4.

	June					
S	M	T	W	T	F	S
~~1~~	~~2~~	~~3~~	~~4~~	~~5~~	~~6~~	~~7~~
~~8~~	~~9~~	~~10~~	~~11~~	~~12~~	~~13~~	~~14~~
~~15~~	~~16~~	~~17~~	~~18~~	~~19~~	~~20~~	~~21~~
~~22~~	~~23~~	24	25	26	27	28
29	30					

5. Graph $\frac{3}{5}$ on a number line

Write the remaining amount as **a.** a mixed number and **b.** an improper fraction.

6. Isabella brought 2 boxes of doughnuts to a meeting. The figure shows the remaining amount of doughnuts.

7. Graph $3\frac{1}{4}$ on a number line.

8. Change $1\frac{3}{5}$ to an improper fraction.

9. Change $\frac{4}{3}$ to a mixed number.

Applications

Solve.

10. In a class of 35 students, 6 students received As on a mathematics exam. What fraction of students received an A? What fraction of students did not receive an A?

11. A certain brand of plain bagels has 146 calories per bagel. 115 calories come from the carbohydrates in the bagel. What fraction of the calories is from carbohydrates?

Writing & Thinking

12. In your own words, list the parts of a fraction and briefly describe the purpose of each part.

13. Show and explain, using diagrams and words, why $2\frac{3}{5} = \frac{13}{5}$.

2.2 Multiplication with Fractions

Multiplying Fractions
1. _____
2. _____

$$\frac{a}{b} \cdot \frac{c}{d} = \underline{} \quad (b, d \neq \underline{})$$

For example, _____

PROCEDURE

Commutative Property of Multiplication
The order of the fractions being multiplied can be _____

$$\frac{a}{b} \cdot \frac{c}{d} = \underline{} \quad (b, d \neq \underline{})$$

For example, _____

PROPERTIES

Associative Property of Multiplication
The grouping of the fractions being multiplied can be _____

$$\left(\frac{a}{b} \cdot \frac{c}{d}\right) \cdot \frac{e}{f} = \underline{} \quad (b, d, f \neq \underline{})$$

For example, _____

PROPERTIES

1. A fraction is reduced to lowest terms if the numerator and denominator _____

Reducing a Fraction to Lowest Terms
1. Factor the _____
2. Use the fact that _____

Note: Reduced fractions may be improper fractions.

PROCEDURE

Watch and Work

Watch the video for Example 11 in the software and follow along in the space provided.

Example 11 Multiplying and Reducing Using Prime Factors

Multiply and reduce to lowest terms: $\dfrac{17}{50} \cdot \dfrac{25}{34} \cdot 8$

Solution

Now You Try It!

Use the space provided to work out the solution to the next example.

Example A Multiplying and Reducing Using Prime Factors

Multiply and reduce to lowest terms: $16 \cdot \dfrac{12}{100} \cdot \dfrac{5}{36}$

2.2 Exercises

Concept Check

True/False. Determine whether each statement is true or false. If a statement is false, explain how it can be changed so the statement will be true. (**Note:** There may be more than one acceptable change.)

1. Multiplication can be used to find $\frac{1}{2}$ of $\frac{2}{9}$.

2. $\frac{3}{4} \cdot \frac{9}{10} = \frac{27}{40}$

3. The statement $\frac{1}{3} \cdot \frac{2}{5} = \frac{2}{5} \cdot \frac{1}{3}$ is an example of the associative property of multiplication.

4. The number 1 is always a factor of the numerator and the denominator.

Practice

Multiply and reduce to lowest terms. (**Hint:** Factor before multiplying.)

5. $\frac{1}{3} \cdot \frac{3}{4}$

6. $\frac{2}{3} \cdot \frac{4}{3}$

7. $\frac{5}{16} \cdot \frac{16}{15}$

8. $\frac{9}{10} \cdot \frac{35}{40} \cdot \frac{25}{15}$

Applications

Solve.

9. A recipe calls for $\frac{3}{4}$ cups of flour. How much flour should be used if only half of the recipe is to be made?

10. A study showed that $\frac{3}{5}$ of the students in an elementary school were left-handed. If the school had an enrollment of 600 students, how many were left-handed?

Writing & Thinking

11. If two fractions are between 0 and 1, can their product be more than 1? Explain.

12. Explain the process of multiplying two fractions. Give an example of a product that cannot be reduced.

2.3 Division with Fractions

Reciprocals

The reciprocal of $\frac{a}{b}$ is _____ . The product of a nonzero number and its reciprocal is _____

$$\frac{a}{b} \cdot \frac{b}{a} = 1$$

Note: $0 = \frac{0}{1}$, but $\frac{1}{0}$ _____

DEFINITION

Dividing Fractions

To divide by any nonzero number, _____

$$\frac{a}{b} \div \frac{c}{d} = $$ _____

For example, _____

PROCEDURE

▶ Watch and Work

Watch the video for Example 5 in the software and follow along in the space provided.

Example 5 Dividing and Reducing Fractions

Divide and reduce to lowest terms: $\frac{16}{27} \div \frac{8}{9}$

Solution

✏️ Now You Try It!

Use the space provided to work out the solution to the next example.

Example A Dividing and Reducing Fractions

Divide and reduce to lowest terms: $\dfrac{10}{21} \div \dfrac{5}{14}$

2.3 Exercises

Concept Check

True/False. Determine whether each statement is true or false. If a statement is false, explain how it can be changed so the statement will be true. (**Note:** There may be more than one acceptable change.)

1. The reciprocal of 1 is undefined.

2. The product of a nonzero number and its reciprocal is undefined.

3. The reciprocal of 12 is $\frac{12}{1}$.

4. The result of $\frac{1}{3} \div \frac{1}{6}$ is 2.

Practice

Divide and reduce to lowest terms.

5. $\frac{2}{3} \div \frac{3}{4}$

6. $0 \div \frac{5}{6}$

7. $\frac{5}{6} \div 0$

8. $\frac{14}{15} \div \frac{21}{25}$

Applications

Solve.

9. The floor of the Atlantic Ocean is spreading apart at an average rate of $\frac{3}{50}$ of a meter per year. How long will it take for the ocean floor to spread 12 meters?

10. An airplane is carrying 180 passengers. This is $\frac{9}{10}$ of the capacity of the airplane.

    a. Is the capacity of the airplane more or less than 180?

    b. If you were to multiply 180 times $\frac{9}{10}$, would the product be more or less than 180?

    c. What is the capacity of the airplane?

Writing & Thinking

11. Explain why the number 0 has no reciprocal.

12. Is division a commutative operation? Explain briefly and give three examples using fractions to help justify your answer.

Name: _____ Date: _____   **59**

2.4 Multiplication and Division with Mixed Numbers

> **Multiplying Mixed Numbers**
> 1. Change each mixed number to _____
>
> 2. Factor the numerator and denominator of _____
>
> 3. Change the answer to _____
>
> **PROCEDURE**

1. The area of a triangle is $\frac{1}{2}$ _____

> **Dividing with Mixed Numbers**
> 1. Change each mixed number to _____
>
> 2. _____
>
> 3. _____
>
> **PROCEDURE**

▶ Watch and Work

Watch the video for Example 10 in the software and follow along in the space provided.

Example 10   Dividing and Reducing Mixed Numbers

Divide and reduce to lowest terms: $3\frac{1}{4} \div 7\frac{4}{5}$

Solution

✏️ Now You Try It!

Use the space provided to work out the solution to the next example.

Example A Dividing and Reducing Mixed Numbers

Divide and reduce to lowest terms: $2\frac{2}{5} \div 3\frac{3}{4}$

2.4 Exercises

Concept Check

True/False. Determine whether each statement is true or false. If a statement is false, explain how it can be changed so the statement will be true. (**Note:** There may be more than one acceptable change.)

1. When multiplying or dividing with mixed numbers, the answer should always be simplified, if possible.

2. Multiplication or division with mixed numbers can be accomplished by changing the mixed numbers to improper fractions.

3. The mixed number $4\frac{1}{5}$ is equal to $\frac{9}{5}$.

4. The reciprocal of $7\frac{2}{5}$ is $\frac{5}{37}$.

Practice

Multiply and reduce to lowest terms. Write your answer in mixed number form.

5. $\dfrac{2}{3} \cdot 3\dfrac{1}{4}$

6. $12\dfrac{1}{2} \cdot 3\dfrac{1}{3}$

Divide and reduce to lowest terms. Write your answer in mixed number form.

7. $3\dfrac{1}{2} \div \dfrac{7}{8}$

8. $7\dfrac{1}{5} \div 3$

Applications

Solve.

9. You are planning a trip of 615 miles (round trip), and you know that your car gets an average of $27\frac{1}{3}$ miles per gallon of gas. You also know that your gas tank holds $15\frac{1}{2}$ gallons of gas.

   a. How many gallons of gas will you use on this trip?

   b. If the gas you buy costs $4 per gallon, how much should you plan to spend on this trip for gas?

10. A right triangle is a triangle with one right angle (measure 90°). The two sides that form the right angle are called legs, and they are perpendicular to each other. The longest side is called the hypotenuse. The legs can be treated as the base and height of the triangle. Find the area of the right triangle shown here.

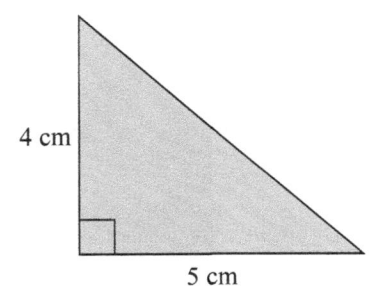

    4 cm

    5 cm

2.4 Exercises

Writing & Thinking

11. Suppose the product of $5\frac{7}{10}$ and some other number is $10\frac{1}{2}$. Answer the following questions without doing any calculations.

    a. Do you think that this other number is more than 1 or less than 1? Why?

    b. Find the other number.

12. Compare and contrast multiplying two mixed numbers and dividing two mixed numbers.

Name: _____ Date: _____ **63**

2.5 Least Common Multiple (LCM)

The **multiples** of a number are _____

> ### Least Common Multiple (LCM)
> The **least common multiple (LCM)** of two (or more) counting numbers is _____
> _____
>
> **DEFINITION**

> ### Finding the LCM of a Set of Counting Numbers
> 1. Find the _____
> 2. List the _____
> 3. Find the product of these primes using each _____
>   _____
>
> **PROCEDURE**

▶ Watch and Work

Watch the video for Example 4 in the software and follow along in the space provided.

Example 4 Finding the Least Common Multiple (LCM)

Find the LCM of 27, 30, and 42.

Solution

2.5 Exercises

✏️ Now You Try It!

Use the space provided to work out the solution to the next example.

Example A Finding the Least Common Multiple (LCM)

Find the LCM of 36, 45, and 60.

Finding Equivalent Fractions

To find a fraction equivalent to $\frac{a}{b}$, multiply the _____

_____.

$$\frac{a}{b} = \frac{a}{b} \cdot \underline{}$$

For example, _____

PROCEDURE

2.5 Exercises

Concept Check

True/False. Determine whether each statement is true or false. If a statement is false, explain how it can be changed so the statement will be true. (**Note:** There may be more than one acceptable change.)

1. The LCM of 15 and 25 is 50.

2. The first five multiples of 9 are 9, 18, 27, 36, and 45.

3. The first five multiples of 4 are 4, 8, 12, 20, and 24.

4. When given larger numbers, the most efficient way to find the LCM is to use the prime factorization method.

Practice

Find the LCM of each set of numbers.

5. 6, 10

6. 3, 4, 8

7. For 14, 35, and 49, **a.** find the LCM and **b.** state how many times each number divides into the LCM.

For each equation, find the missing numerator that will make the fractions equivalent.

8. $\dfrac{5}{8} = \dfrac{?}{24}$

9. $\dfrac{5}{12} = \dfrac{?}{108}$

2.5 Exercises

Applications

Solve.

10. Three security guards meet at the front gate for coffee before they walk around inspecting buildings at a manufacturing plant. The guards take 15, 20, and 30 minutes, respectively, for the inspection trip.

    a. If they start at the same time, in how many minutes will they meet again at the front gate for coffee?

    b. How many trips will each guard have made?

11. A fruit production company has three packaging facilities, each of which uses different-sized boxes as follows: 24 pieces/box, 36 pieces/box, and 45 pieces/box.

    a. Assuming that the truck provides the same quantity of uniformly-sized pieces of fruit to all three packaging facilities, what is the minimum number of pieces of fruit that will be delivered so that no fruit will be left over?

    b. How many boxes will each facility package?

Writing & Thinking

12. Explain, in your own words, why each number in a set divides evenly into the LCM of that set of numbers.

13. Explain why simply multiplying two numbers together will not necessarily find the LCM of those numbers. Give an example of when it would find the LCM and an example when it would not.

2.6 Addition and Subtraction with Fractions

Adding Fractions with the Same Denominator
1. _____
2. _____
3. _____

$\frac{a}{b} + \frac{c}{b} =$ _____

For example, _____

PROCEDURE

Adding Fractions with Different Denominators
1. Find the _____
2. Change each fraction into _____
3. _____
4. _____

PROCEDURE

Commutative Property of Addition
The order of the fractions being added can be _____

$$\frac{a}{b} + \frac{c}{d} = \text{_____}$$

For example, _____

PROPERTIES

Associative Property of Addition
The grouping of the fractions being added can be _____

$$\frac{a}{b} + \left(\frac{c}{d} + \frac{e}{f} \right) = \text{_____}$$

For example, _____

PROPERTIES

2.6 Addition and Subtraction with Fractions

> **Subtracting Fractions with the Same Denominator**
> 1. _____
> 2. _____ $\dfrac{a}{b} - \dfrac{c}{b} =$ _____
> 3. _____
>
> For example, _____
>
> **PROCEDURE**

> **Subtracting Fractions with Different Denominators**
> 1. Find the _____
> 2. Change each fraction into _____
> 3. _____
> 4. _____
>
> **PROCEDURE**

▶ Watch and Work

Watch the video for Example 11 in the software and follow along in the space provided.

Example 11 Subtracting Fractions with Different Denominators

Subtract: $1 - \dfrac{5}{8}$

Solution

Now You Try It!

Use the space provided to work out the solution to the next example.

Example A Subtracting Fractions with Different Denominators

Subtract: $3 - \dfrac{5}{12}$

2.6 Exercises

Concept Check

True/False. Determine whether each statement is true or false. If a statement is false, explain how it can be changed so the statement will be true. (**Note:** There may be more than one acceptable change.)

1. The final step in adding fractions is to reduce, if possible.

2. The process for finding the LCD is the same as the process for finding the LCM.

3. LCD represents the Least Common Digit.

4. When subtracting fractions, simply subtract the numerators and the denominators.

5. Subtraction of fractions requires that the fractions have the same denominators.

Practice

Add and reduce to lowest terms.

6. $\dfrac{3}{25} + \dfrac{12}{25}$

7. $\dfrac{2}{7} + \dfrac{4}{21} + \dfrac{1}{3}$

Subtract and reduce to lowest terms.

8. $\dfrac{7}{8} - \dfrac{5}{8}$

10. $2 - \dfrac{9}{16}$

9. $\dfrac{9}{14} - \dfrac{2}{21}$

Applications

Solve.

11. Three pieces of mail weigh $\frac{1}{2}$ ounce, $\frac{1}{5}$ ounce, and $\frac{3}{10}$ ounce. What is the total weight of the letters?

12. A recipe calls for the following spices: $\frac{1}{2}$ teaspoon of turmeric, $\frac{1}{4}$ teaspoon of ginger, and $\frac{1}{8}$ teaspoon of cumin. What is the total quantity of these three spices?

Writing & Thinking

13. Explain how finding the LCM relates to LCDs.

14. Give an example of a situation where you might add or subtract fractions (other than in class).

2.7 Addition and Subtraction with Mixed Numbers

Adding Mixed Numbers
1. _____
2. _____
3. Write the answer as _____

PROCEDURE

Subtracting Mixed Numbers
1. _____
2. _____

PROCEDURE

Subtracting Mixed Numbers by Borrowing
1. "Borrow" 1 from _____
2. Add this 1 to _____ (This will always result in _____ _____)
3. _____

PROCEDURE

▶ Watch and Work

Watch the video for Example 8 in the software and follow along in the space provided.

Example 8 Subtracting Mixed Numbers by Borrowing

Subtract: $4\frac{2}{9} - 1\frac{5}{9}$

Solution

Now You Try It!

Use the space provided to work out the solution to the next example.

Example A Subtracting Mixed Numbers by Borrowing

Subtract: $10\frac{1}{8} - 3\frac{5}{8}$

2.7 Exercises

Concept Check

True/False. Determine whether each statement is true or false. If a statement is false, explain how it can be changed so the statement will be true. (**Note:** There may be more than one acceptable change.)

1. $3\frac{1}{5} + 5\frac{1}{2} = 8\frac{7}{10}$

2. When adding (or subtracting) mixed numbers, the final answer should be written as a mixed number.

3. LCDs are not required when adding or subtracting mixed numbers.

4. $12 - 5\frac{1}{3} = 7\frac{1}{3}$

Practice

Add and reduce to lowest terms. Write your answer in mixed number form.

5. $4\dfrac{1}{2} + 3\dfrac{1}{6}$

6. $7\dfrac{3}{5} + 2\dfrac{1}{8}$

Subtract and reduce to lowest terms. Write your answer in mixed number form.

7. $7\dfrac{9}{10}$
   $-3\dfrac{3}{10}$

8. $4\dfrac{9}{16}$
   $-2\dfrac{7}{8}$

Applications

Solve.

9. A bus trip is made in three parts. The first part takes $2\frac{1}{3}$ hours, the second part takes $2\frac{1}{2}$ hours, and the third part takes $3\frac{3}{4}$ hours. How long does the entire trip take?

10. On average, the air that we inhale includes $1\frac{1}{4}$ parts water and the air we exhale includes $5\frac{9}{10}$ parts water. How many more parts water are in the exhaled air?

Writing & Thinking

11. In subtracting with mixed numbers explain why fractional parts should be subtracted before the whole numbers.

2.8 Comparisons and Order of Operations with Fractions

For any two numbers on a number line, the smaller number is to the _____ of the larger number.

Comparing Two Fractions

1. Find the _____

2. Change each fraction into _____

3. _____

PROCEDURE

Rules for Order of Operations

1. Simplify within grouping symbols, such as _____
   _____

2. Evaluate any _____

3. Moving from left to right, perform any _____
   _____

4. Moving from left to right, perform any _____

PROCEDURE

▶ Watch and Work

Watch the video for Example 6 in the software and follow along in the space provided.

Example 6 Using the Order of Operations with Fractions

Simplify: $3\dfrac{2}{5} \div \left(\dfrac{1}{4} + \dfrac{3}{5}\right)$

Solution

2.8 Comparisons and Order of Operations with Fractions

✏️ Now You Try It!

Use the space provided to work out the solution to the next example.

Example A Using the Order of Operations with Fractions

Simplify: $\left(\dfrac{5}{12} + \dfrac{2}{3}\right) \div 4\dfrac{1}{3}$

2.8 Exercises

Concept Check

True/False. Determine whether each statement is true or false. If a statement is false, explain how it can be changed so the statement will be true. (**Note:** There may be more than one acceptable change.)

1. The rules for order of operations are the same for fractions, mixed numbers, and whole numbers.

2. According to the rules for order of operations, multiplication occurs before division.

3. An average is found by adding all the numbers in the set and then dividing by the quantity of numbers in the set.

Practice

For each pair of fractions, determine which fraction is larger and by how much it is larger.

4. $\frac{2}{3}, \frac{3}{4}$

5. $\frac{4}{5}, \frac{17}{20}$

6. Arrange $\frac{2}{3}, \frac{3}{4}, \frac{5}{8}$ in order from smallest to largest. Then find the difference between the largest and smallest fractions.

2.8 Exercises

Simplify.

7. $\dfrac{1}{2} \div \dfrac{7}{8} + \dfrac{1}{7} \cdot \dfrac{2}{3}$

8. $\dfrac{1}{2} \div \dfrac{2}{3} + \left(\dfrac{1}{3}\right)^2$

9. Find the average of the numbers $\dfrac{5}{6}, \dfrac{1}{15},$ and $\dfrac{17}{30}$.

Applications

Solve.

10. Two painters paint $76\frac{1}{2}$ feet of fencing in one day. The first painter works for $2\frac{1}{2}$ hours and the second works for $5\frac{3}{5}$ hours. How many feet of fencing are painted in each hour? (Hint: Add the number of hours then divide the length of fencing by this sum.)

11. An art book has 40 two-sided pages of pictures, each of which is $\frac{1}{32}$ inch thick. Each two-sided page is protected by a $\frac{1}{80}$ inch thick piece of paper. Each side of the book is bound by a $\frac{1}{6}$ inch cover. What is the total thickness of the book?

Writing & Thinking

12. **a.** If two fractions are between 0 and 1, can their sum be more than 1? Explain.

    b. If two fractions are between 0 and 1, can their sum be more than 1? Explain.

Name: Date: **79**

Chapter 2 Project

On a Budget
An activity to demonstrate the use of fractions in real life.

Samantha's friend Meghan is always traveling to exotic locations when she goes on vacation. One day, Samantha asked Meghan how she was able to afford it. Meghan told her it was simple: she makes a budget and sets aside a portion of her income each month for a vacation. Meghan told Samantha that they could meet at her house for dinner on Saturday and she would be glad to show her how to make a budget.

The table below shows the amount Meghan spends on each category in her budget. Assume Meghan makes $4400 a month (after taxes) and answer the following questions. Fractions should be written in reduced form.

Meghan's Budget

Category	Amount	Category	Amount
Rent	$1100	Car Loan	$528
Groceries	$352	IRA	$264
Utilities	$220	Charity	$440
Dining Out	$264	Savings	$275
Vacation	$440	Car Insurance	$176
Gas	$88	Phone	$110

1. Samantha earns a different amount per month and wants to compare her spending to Meghan's, so she decides to convert all of Meghan's spending to fractions. Find the fractional values for Meghan's categories by dividing each amount by her monthly income and reducing the fractions to lowest terms.

Meghan's Budget

Category	Amount	Category	Amount
Rent		Car Loan	
Groceries		IRA	
Utilities		Charity	
Dining Out		Savings	
Vacation		Car Insurance	
Gas		Phone	

2. Find the sum of the fractions in the budget. Begin by adding the fractions that have like denominators, then add the fractions by finding a common denominator.

3. Samantha tells Meghan that her budget must be incomplete. Meghan informs her that the remaining is what she spends on miscellaneous items like cleaning supplies, toiletries, and clothes. What fraction of her budget does Meghan spend on miscellaneous items?

4. Samantha determines that she spends $\frac{1}{10}$ of her income on groceries and $\frac{1}{8}$ dining out. What fraction of her income in spent on food?

5. What fraction of Megan's income is spent on food?

6. Find the difference between the fractional amount that Meghan and Samantha spend for food. Who uses the largest fractional part of their income for food?

7. Add the fractional values for the car loan, car insurance, and gas to determine the fractional amount of Meghan's income that is set aside for her car.

8. Samantha doesn't have a car payment, but she has to pay for parking at her apartment. She has determined that she spends $\frac{1}{50}$ of her monthly income on car insurance, $\frac{3}{50}$ on parking, and $\frac{1}{25}$ on gas. What fractional part of her income is used for her car?

9. Find the difference between the fractional amount that Meghan and Samantha spend on their cars. Who uses the largest fractional part of their income for car expenses?

10. Meghan plans to move into a new apartment with a roommate and travel more often. She needs to revise her budget accordingly.

Meghan's Budget

Category	Amount	Category	Amount
Rent		Car Loan	
Groceries		IRA	
Utilities		Charity	
Dining Out		Savings	
Vacation		Car Insurance	
Gas		Phone	

a. Halve the first three fractions in the first column. Since Megan will have a roommate and will be gone more, she expects to spend half as much on rent, groceries, and utilities. Place the new values in the new table above.

b. Double the last three fractions in the first column. Since Megan will be traveling, she expects to spend twice as much for dinning out, vacation, and gas. Place the new values in the table.

c. Find the total fractional portion budgeted for the new budget plan if the second column remains the same.

d. Would Meghan be able to afford this new budget on her current income? In other words, do the new fractions in the table add up to a fraction less than 1?

e. If the new budget in part d. is over budget, what fraction is it over? If the new budget is under budget, what fractional portion is left over for miscellaneous items?

Chapter 2 Project

What's Cookin', Good Lookin'?

An activity to demonstrate the use of fractions in real life.

You are baking shortbread cookies for two events: your mother's birthday and your mathlete's celebration party. You have your grandfather Giuseppe's famous recipe and want to make the exact amount of cookies needed for both parties with one batch of dough, meaning you must adjust the size of the original recipe.

Giuseppe's Famous Shortbread Cookies

Ingredient	Amount Needed
Butter	1 cup
Sugar	$\frac{1}{2}$ cup
Vanilla Extract	1 teaspoon
Salt	1 teaspoon
Flour	2 cups

Makes 36 cookies

No extra directions are given, typical of Grandpa, huh?

1. Eighteen people are coming to your mother's birthday and 9 people are coming to the celebration party. How many total cookies do you need, assuming each person eats one cookie?

2. First, let's determine how to adjust the recipe to make cookies for 18 people.

   a. What fraction of the recipe do you need to make 18 cookies? (**Hint**: Place the part over the whole number of cookies the recipe makes and reduce.)

   b. Multiply each of the ingredients by the fraction you found in part a.

3. Next, let's determine how to adjust the recipe to make cookies for 9 people.

   a. What fraction of the recipe do you need to make 9 cookies?

   b. Multiply each of the ingredients by the fraction you found in part a.

4. Find the sum of the individual ingredients from Problems 2 and 3 to create one new recipe. Rewrite any improper fractions as mixed numbers.

5. Now you have an adjusted recipe to make the exact number of cookies needed. You have the following measuring cups and spoons: 1 cup, $\frac{1}{2}$ cup, $\frac{1}{4}$ cup, 1 teaspoon, $\frac{1}{2}$ teaspoon, $\frac{1}{4}$ teaspoon, and 1 tablespoon.

   a. How would you measure out $\frac{3}{4}$ cup of an ingredient with the given cups?

   b. Is it possible to measure out all of your ingredients with the given cups and spoons? If not, which ingredients?

   c. Find at least two pairs of fractions that add to $\frac{3}{8}$.

   d. If $\frac{1}{4}$ cup is equal to 4 tablespoons, how many tablespoons would be in $\frac{1}{8}$ cup?

   e. How could you measure out $\frac{3}{8}$ cup of an ingredient with this new knowledge?

   f. List at least two ways you could measure out the needed amount of flour.

6. Your sister gifted you with some local artisanal pecan flour, and you want substitute some of the flour in the recipe with the pecan flour with a one-to-one substitution. Using the available measuring cups, list three combinations of measurements that you could use to combine the normal flour and pecan flour and still measure out the total needed amount of flour.

7. Is it necessary to create a new adjusted recipe each time you want to make a certain amount of cookies? What are some additional strategies you could use to make a certain amount of cookies for a party?

8. Perform an internet search for recipes for your top three favorite cookies.

   a. State at least three ingredients that are commonly found in these recipes that are not included in the shortbread cookie recipe.

   b. Do you think any of these additional common baking ingredients are difficult to halve? Are any of these easy to double?

9. When doubling or halving a recipe, some common baking ingredients should not be doubled or halved. Perform an internet search and explain in your own words why that is true.

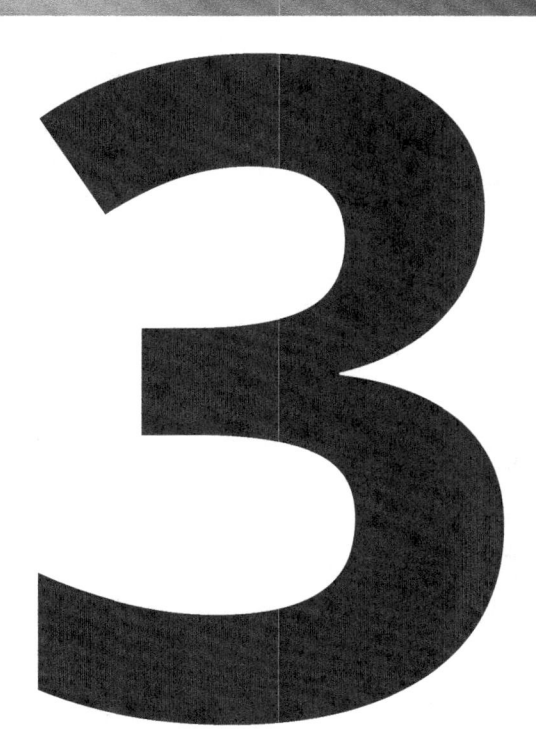

CHAPTER 3

Decimal Numbers

3.1 Introduction to Decimal Numbers

3.2 Addition and Subtraction with Decimal Numbers

3.3 Multiplication with Decimal Numbers

3.4 Division with Decimal Numbers

3.5 Estimating and Order of Operations with Decimal Numbers

3.6 Decimal Numbers and Fractions

CHAPTER 3 PROJECTS
What Would You Weigh on the Moon?

A Trip to the Grocery Store

Connections

Many people use decimal numbers on a daily basis to discuss sports. Batting averages in baseball and save percentages in hockey are both calculated to the nearest thousandth. Quarterback ratings in football and average points per game in basketball are both calculated to the nearest tenth. Decimal numbers are also used to record many of the world's sport records. For example, for the 100-meter freestyle, the men's swimming record is 44.84 seconds by Kyle Chalmers of Russia, and the women's swimming record is 51.71 seconds by Sarah Sjöström of Sweden.

During the NBA 2021–2022 regular season, the top three players based on points per game were Giannis Antetokounmpo of the Milwaukee Bucks, Luka Doncic of the Dallas Mavericks, and Nikola Jokic of the Denver Nuggets. Antetokounmpo and Doncic tied with 31.7 points per game, followed by Jokic with 31.0 points per game. During the season, Antetokounmpo played 67 games, Doncic played 65 games, and Jokic played 74 games. How would you determine how many total points each player scored during the 2021–2022 regular season? Which player scored the most points during the season?

Source: "Season Leaders|Stats", NBA, Accessed August 9, 2022, www.nba.com/stats/players.

Name: _____ Date: _____   **85**

3.1 Introduction to Decimal Numbers

> ### Reading or Writing a Decimal Number
> 1. Read (or write) the _____
> 2. Read (or write) the _____
> 3. Read (or write) the _____
>
>   Then, name the fraction part with the _____
>
>   PROCEDURE

> ### Comparing Two Decimal Numbers
> 1. Moving left to right, compare digits _____
> 2. When one compared digit is larger, then the _____
>
>   PROCEDURE

▶ Watch and Work

Watch the video for Example 6 in the software and follow along in the space provided.

Example 6 Comparing Decimal Numbers

Arrange the following three decimal numbers in order from smallest to largest: 6.67, 5.14, 6.28. Then, graph them on a number line.

Solution

Now You Try It!

Use the space provided to work out the solution to the next example.

Example A Comparing Decimal Numbers

Arrange the following three decimal numbers in order from smallest to largest: 2.6, 2.06, 2.46. Then, graph them on a number line.

Rounding Rule for Decimal Numbers

1. Look at the single digit one place value to the right of the digit in the place of desired accuracy.

   a. **If this digit is less than 5,** _____ and replace all digits to the right with zeros. All digits to the _____

   b. **If this digit is 5 or greater,** _____ and replace all digits to the right with zeros. All digits to the _____

   _____. Then, the 9 is replaced by 0 and the _____

2. Zeros to the right of the place of accuracy that are also to the right of the _____
   _____

PROCEDURE

3.1 Exercises

Concept Check

True/False. Determine whether each statement is true or false. If a statement is false, explain how it can be changed so the statement will be true. (**Note:** There may be more than one acceptable change.)

1. Two hundred thousand, four hundred six and twelve hundredths can be written as 200,406.12.

2. 92.586 is greater than 92.6.

3. On a number line, any number to the left of another number is larger than that other number.

4. When a decimal number is rounded, all numbers to the right of the place of accuracy become zeros in the final answer.

Practice

5. Write $2\frac{57}{100}$ in decimal notation.

6. Write 20.7 in words.

7. Write six and twenty-eight thousandths in decimal notation.

8. Arrange 0.2, 0.26, and 0.17 in order from smallest to largest. Then, graph the numbers on a number line.

3.1 Exercises

Fill in the blanks to correctly complete each statement.

9. Round 3.00652 to the nearest ten-thousandth.

   a. The digit in the ten-thousandths position is ___.

   b. The next digit to the right is ___.

   c. Since ___ is less than 5, leave ___ as it is and replace ___ with 0.

   d. So 3.00652 rounds to _____ to the nearest ten-thousandth.

Applications

In each exercise, write the decimal numbers that are not whole numbers in words.

10. The tallest unicycle ever ridden was 114.8 feet tall, and was ridden by Sam Abrahams (with a safety wire suspended from an overhead crane) for a distance of 28 feet in Pontiac, Michigan, on January 29, 2004.[1]

11. One quart of water weighs approximately 2.0825 pounds.

Writing & Thinking

12. Discuss situations where you think it is particularly appropriate (or necessary) to write numbers in English word form.

13. With a. and b. as examples, explain in your own words how you can tell quickly when one decimal number is larger (or smaller) than another decimal number.

    a. The decimal number 2.765274 is larger than the decimal number 2.763895.
    b. The decimal number 17.345678 is larger than the decimal number 17.345578.

[1] Source: semcycle.biz/record

3.2 Addition and Subtraction with Decimal Numbers

Adding Decimal Numbers
1. Write the _____
2. Keep the _____
3. _____
4. _____, keeping the decimal point in the sum aligned with the other decimal points.

PROCEDURE

▶ Watch and Work

Watch the video for Example 3 in the software and follow along in the space provided.

Example 3 Adding Decimal Numbers

Add: $9 + 4.86 + 37.479 + 0.6$

Solution

✏ Now You Try It!

Use the space provided to work out the solution to the next example.

Example A Adding Decimal Numbers

Add: $23.8 + 4.2567 + 11 + 3.01$

90   3.2   Exercises

> **Subtracting Decimal Numbers**
>
> 1. Write the _____
>
> 2. Keep the _____
>
> 3. Keep digits with the _____
>
> 4. _____, keeping the decimal point in the difference aligned with the other decimal points.
>
> PROCEDURE

3.2 Exercises

Concept Check

True/False. Determine whether each statement is true or false. If a statement is false, explain how it can be changed so the statement will be true. (**Note:** There may be more than one acceptable change.)

1. Vertical alignment is the preferred method of adding decimal numbers because digits with the same place value are easily lined up.

2. It is important to align the decimal points vertically when adding decimal numbers.

3. In subtracting decimal numbers, line up all the last digits vertically.

4. Once decimal points and corresponding digits have been aligned vertically, add or subtract from left to right.

Practice

Add.

5. 42.08
   + 8.005

6. 1.3408
   + 6.539

Subtract.

7. 39.542
   − 28.411

8. 87.1
   − 69.3

Applications

Solve.

9. Mr. Johnson bought the following items at a department store: slacks, $32.50; shoes, $43.75; shirt, $18.60.

   a. How much did he spend?

   b. What was his change if he gave the clerk a $100 bill? (Tax was included in the prices.)

3.2 Exercises

10. David is preparing a four-sided garden plot with unequal sides of 7.5 feet, 26.34 feet, 36.92 feet, and 12.07 feet. How many feet of edging material must he use? (This is the same as finding the perimeter of the plot.)

Writing & Thinking

11. Why is it important that the decimal points and numbers be aligned vertically when adding or subtracting decimals?

12. Suppose that you are given two decimal numbers with 0 as their whole number part.

    a. Explain how the sum might be more than 1.

    b. Explain why the sum cannot be more than 2.

3.3 Multiplication with Decimal Numbers

Multiplying Decimal Numbers

1. Multiply the two _____
2. Count the total number of digits to the _____
   _____
3. Place the decimal point in the product so that the number of digits to the _____
   _____

PROCEDURE

▶ Watch and Work

Watch the video for Example 2 in the software and follow along in the space provided.

Example 2 Multiplying Decimal Numbers

Multiply: $4.35(12.6)$

Solution

✏ Now You Try It!

Use the space provided to work out the solution to the next example.

Example A Multiplying Decimal Numbers

Multiply: $(5.65)(1.14)$

> **Multiplying by Powers of 10 (10, 100, 1000, and so on)**
> 1. Count the number of _____
> 2. Move the decimal point to the _____
>   _____
>
>   Multiplication by 10 moves the decimal point _____
>
>   Multiplication by 100 moves the decimal point _____
>
>   Multiplication by 1000 moves the decimal point _____
>
> And so on.
>
> **PROCEDURE**

3.3 Exercises

Concept Check

True/False. Determine whether each statement is true or false. If a statement is false, explain how it can be changed so the statement will be true. (**Note:** There may be more than one acceptable change.)

1. The decimal points should be aligned vertically when multiplying decimal numbers.

2. The first step in the multiplication of decimal numbers is to multiply the numbers as if they were whole numbers.

3. When multiplying decimal numbers, the answer should have the same number of decimal places as the total number of decimal places in the numbers being multiplied.

4. Multiplying by 100 requires that the decimal point be moved 100 places to the right.

Practice

Multiply.

5. $(0.6)(0.7)$

6. $4(0.1213)$

7. $\begin{array}{r} 0.08 \\ \times\, 0.542 \\ \hline \end{array}$

8. $1000(4.1782)$

Applications

Solve.

9. To buy a car, you can pay $2036.50 in cash, or you can put down $400 and make 18 monthly payments of $104.30. How much would you save by paying cash?

10. The salesperson at a clothing store receives a salary of $375.50 plus a weekly commission of 0.18 times all sales over $850 for the week. How much will that person earn if he sells $3945 that week? (**Hint:** Compute how much over the threshold of $850 the total sales were, and then multiply it by the rate of commission. Then add his salary to that value.)

Writing & Thinking

11. In your own words, discuss the similarities and differences between multiplication with whole numbers and multiplication with decimal numbers.

12. Discuss, briefly, situations in which you might use multiplication with decimal numbers in your daily life.

3.4 Division with Decimal Numbers

Dividing Decimal Numbers

1. Move the decimal point in the _____
2. Move the decimal point in the _____
3. Place the decimal point in the _____
4. Divide just as with whole numbers.

PROCEDURE

Dividing When the Remainder is Not 0

1. Decide how many _____
2. Divide until the quotient has been calculated to _____ _____
3. Using this last digit, round the _____

PROCEDURE

Terminating and Nonterminating Decimal Numbers

If the remainder is eventually 0, the decimal number is _____. For example, _____ _____.

If the remainder is not eventually 0, the decimal number is _____

For example, _____.

DEFINITION

▶ Watch and Work

Watch the video for Example 3 in the software and follow along in the space provided.

Example 3 Dividing Decimal Numbers

Divide (to the nearest tenth). $82.3 \div 2.9$

Solution

✏ Now You Try It!

Use the space provided to work out the solution to the next example.

Example A Dividing Decimal Numbers

Divide (to the nearest tenth). $83.5 \div 5.6$

> ### Dividing a Decimal Number by a Power of 10 (10, 100, 1000, and so on)
> 1. Count the number of _____
> 2. Move the decimal point to the _____
> _____
>
> Division by **10** moves the decimal point _____
>
> Division by **100** moves the decimal point _____
>
> Division by **1000** moves the decimal point _____
>
> And so on.
>
> **PROCEDURE**

3.4 Exercises

Concept Check

True/False. Determine whether each statement is true or false. If a statement is false, explain how it can be changed so the statement will be true. (**Note:** There may be more than one acceptable change.)

1. The first step in division with decimal numbers is to move the decimal point in the divisor to the right so that the divisor is a whole number.

2. Moving the decimal point in a divisor requires that the decimal point also be moved in the dividend.

3. The decimal point should be placed in the quotient before actually dividing.

4. Dividing by a power of 10 involves dividing by 10, 20, 30, 40, etc.

Practice

Divide. Round to the nearest hundredth, if necessary.

5. $4.95 \div 5$

6. $7.336 \div 1.4$

7. $\dfrac{0.1463}{24}$

8. $\dfrac{138.1}{10}$

Applications

Solve.

9. If you bought 6 books for a total price of $142.98, what average amount did you pay per book, including tax?

10. A professor has graded a test of five students, and their scores were 76.4, 100, 84.7, 10.2, and 68.3. What is the average of these five scores?

Writing & Thinking

11. In your own words, discuss the similarities and differences between division with whole numbers and division with decimal numbers.

12. List the steps you would follow in working a division problem with decimal numbers.

3.5 Estimating and Order of Operations with Decimal Numbers

1. We can estimate a sum (or difference) by _____

   _____

▶ Watch and Work

Watch the video for Example 1 in the software and follow along in the space provided.

Example 1 Estimating Sums of Decimal Numbers

Estimate the sum; then find the actual sum.

$$74 + 3.529 + 52.61$$

Solution

✏ Now You Try It!

Use the space provided to work out the solution to the next example.

Example A Estimating Sums of Decimal Numbers

Estimate the sum; then find the actual sum.

$$6.68 + 103 + 21.94$$

2. Estimating products can be done by _____

_____

3. In order to estimate with division, _____

_____

3.5 Exercises

Concept Check

True/False. Determine whether each statement is true or false. If a statement is false, explain how it can be changed so the statement will be true. (**Note:** There may be more than one acceptable change.)

1. An estimate of the sum 71.369 + 49.1 is 120.

2. One way to estimate the product of decimal numbers is to round the numbers to the rightmost nonzero digit before performing the multiplication.

3. An estimate of the quotient 16.469 ÷ 3.87 would be 4.

4. Experience and understanding are needed to decide whether or not a particular answer is reasonably close to an estimate.

5. According to the rules for order of operations, addition and subtraction should be performed before multiplication and division.

Practice

Estimate each answer, then find the actual answer rounded to the nearest hundredth, if necessary.

6. 29.03
  + 3.79

7. 51.21
  − 25.13

8. $(6.3)(1.6)$

9. $3.1\overline{)6.36}$

Simplify.

10. $8.6 \div 2.15 + 3.6 \cdot 20.3$

Applications

Solve.

11. Jim is packing three sculptures in a box for shipping. The weights of the sculptures are 5.63 pounds, 12.4 pounds, and 3 pounds. The shipping materials weigh 17.4 pounds.

    a. Estimate the total weight.

    b. Find the actual weight.

12. Bicycle racer Peter Sagan rode 125.09 miles in 5.35 hours.

    a. Estimate how fast he was riding per hour.

    b. What was his average speed per hour (to the nearest hundredth)?

Writing & Thinking

13. Suppose you are only interested in an approximate answer for a product. Would there be any difference in the products produced by the following two procedures?

    a. First multiply the two numbers as they are and then round the product to the desired place of accuracy.

    b. First round each number to the desired place of accuracy and then multiply the rounded numbers.

    Explain why you think these two procedures would produce the same result or different results.

3.6 Decimal Numbers and Fractions

> **Changing from Decimal Numbers to Fractions**
>
> A decimal number less than 1 (digits are to the right of the decimal point) can be written in fraction form by writing a fraction with
>
> 1. A numerator that consists of the whole number formed by _____
>   _____
>
> 2. A denominator that is the power of _____
>
>   (For example, a denominator of 100 corresponds to _____.)
>
> PROCEDURE

1. _____ numbers can be **repeating** or **nonrepeating**. A **repeating decimal** has a repeating pattern to its digits.

2. Every fraction with a whole number numerator and nonzero denominator is _____.
   Such numbers are called _____.

▶ Watch and Work

Watch the video for Example 8 in the software and follow along in the space provided.

Example 8 Simplifying Expressions with Decimals and Fractions

Find the sum $10\frac{1}{2} + 7.32 + 5\frac{3}{5}$ in decimal form.

Solution

Now You Try It!

Use the space provided to work out the solution to the next example.

Example A Simplifying Expressions with Decimals and Fractions

Find the sum $2.88 + \frac{1}{4} + 13\frac{9}{10}$ in decimal form.

3.6 Exercises

Concept Check

True/False. Determine whether each statement is true or false. If a statement is false, explain how it can be changed so the statement will be true. (**Note:** There may be more than one acceptable change.)

1. When a decimal number is changed to a fraction, the denominator will be the power of 10 that names the rightmost digit of the decimal number.

2. When a decimal number is changed to a fraction, the numerator can be determined by using the whole number that is formed by all the digits of the decimal number.

3. Fractions can always be converted to decimal form without losing accuracy.

4. In decimal form, $\frac{1}{3}$ is repeating and nonterminating.

Practice

Change each decimal number to a fraction or mixed number in lowest terms.

5. 0.18

6. 2.75

Change each fraction to a decimal number rounded to the nearest hundredth.

7. $\dfrac{20}{3}$

8. $\dfrac{40}{9}$

Simplify the expression by first writing all of the numbers in decimal form. Round to the nearest hundredth, if necessary.

9. $\dfrac{1}{4} + 0.25 + \dfrac{1}{5}$

10. Arrange $0.76, \dfrac{3}{4}, \dfrac{7}{10}$ in order from smallest to largest.

Applications

Solve.

11. A rectangle has a length of 6.4 inches and a width that is $\frac{2}{5}$ times the length. Find the perimeter of the rectangle.

12. A loaf of bread weighs 21.6 ounces. Mauricio cut off a third of the loaf to save for later and then cut the remaining portion into 16 equal slices. What was the weight of each slice of the 16 slices he cut?

Writing & Thinking

13. Describe the process used to change a terminating decimal number to a fraction.

14. List 2 different ways to solve this problem: $\frac{1}{2} + 3.67 - \frac{1}{8}$. State which method you prefer and why.

Chapter 3 Project

What Would You Weigh on the Moon?
An activity to demonstrate the use of whole numbers in real life.

The following table contains the surface gravity of each planet in the same solar system as the Earth, as wells Earth's moon, and the sun. The acceleration due to gravity g at the surface of a planet is given by the formula

$$g = \frac{GM}{R^2},$$

where M is the mass of the planet, R is its radius, and G is the gravitational constant. From the formula you can see that a planet with a larger mass M will have a greater value for surface gravity. Also the larger the radius R of the planet, the smaller the surface gravity.

If you look at different sources, you may find that surface gravity varies slightly from one source to another due to different values for the radius of some planets, especially the gas giants: Jupiter, Saturn, Uranus, and Neptune.

Planet	Surface Gravity (m/s²)	Relative Surface Gravity	Fractional Equivalent
Earth	9.78	1.00	
Jupiter	23.10	2.36	
Mars	3.72		
Mercury	3.78		
Moon	1.62		
Neptune	11.15		
Saturn	9.05		
Sun	274.00		
Uranus	8.69		
Venus	9.07		

1. Compare the surface gravity of each planet or celestial body to the surface gravity of the Earth by dividing each planet's surface gravity by that of the Earth's, as listed in the table above. (This is referred to as **relative surface gravity**.) Round your answer to the nearest hundredth and place your results in the third column of the table. The values for Earth and Jupiter have been done for you. (**Note:** Comparing Earth to itself results in a value of 1.)

2. For Jupiter, the relative surface gravity value of 2.36 means that the gravity on Jupiter is 2.36 times that of Earth; therefore, your weight on Jupiter would be approximately 2.36 times your weight on Earth. (Although mass is a constant and doesn't change regardless of what planet you are on, your weight depends on the pull of gravity). Explain what the relative surface gravity value means for Mars.

3. Calculate your weight on the moon by taking your present weight (in kg or pounds) and multiplying it by the moon's relative surface gravity.

4. Approximately how many times larger is the surface gravity of the sun compared to that of Mars? Round to the nearest whole number.

5. Convert each value in column three to a mixed number and place the result in column four. Be sure to reduce all fractions to lowest terms.

   a. Which is larger, the relative surface gravity of Mercury or $\frac{2}{5}$?

   b. Which is smaller, the relative surface gravity of the moon or $\frac{4}{25}$?

   c. Write the fractional equivalent of Jupiter's relative surface gravity as an improper fraction in lowest terms.

Chapter 3 Project

A Trip to The Grocery Store

An activity to demonstrate the use of decimals in real life.

Decimals can be encountered in many areas of our daily lives. When we order food at a restaurant or go to the gas station, we see numbers written and expressed to the nearest hundredth. The grocery store is the same; we see prices for items, prices per unit, sale stickers, and so on, and we need to be able to find the best prices to get the most for our money.

For your upcoming trip to the grocery store, and suppose you have made a full list of items that you will need to purchase. In each question, you will be evaluating the prices of each item on your list using various operations with decimals.

1. The first three items you need to buy are a gallon of milk, 2 cartons of eggs, and a stick of butter. You have found that $3.49 is the best price for milk, $2.57 is the best price for one carton of eggs, and $0.93 is the best price for a stick of butter. What is the total cost of these items?

2. For canned beans, the grocery store is running a promotional deal where you will receive 50 cents off each canned good as long as you buy 10 of them. If the price of one can of beans is $1.34, what is the total price for 10 cans of beans? Explain how you found your answer.

3. When it comes to buying a loaf of bread, you have several options. To find the best deal, in this case, we need to find the unit price. The unit price is the cost per one item in a larger group. In other words, the unit price of the bread would be the cost per each slice of bread as opposed to the cost for the entire loaf. Assume each slice is the same size.

    Bread Brand A has 25 slices and costs $2.35.
    Bread Brand B has 18 slices and costs $1.95.

    a. What is the unit price for a slice of Brand A's bread? Round to the nearest hundredth.

    b. What is the unit price for a slice of Brand B's bread? Round to the nearest hundredth.

    c. Which brand has the better deal per slice of bread? How do you know?

4. The store has three brands of paper towels for sale, with varying prices and amounts of individual sheets. Assume each sheet has the same dimensions.

    Brand C is $12.90 for 400 sheets.
    Brand D is $8.36 for 500 sheets.
    Brand E is $9.30 for 775 sheets.

    a. Calculate the unit price for each brand per sheet. Round each to the nearest hundredth.

    b. During the weekend, the grocery store is running a sale on paper towels. If Brand C is on sale for a discount of 20%, is it now cheaper than Brand E, by unit price? Explain your answer.

    c. What total price would Brand D need to be to have the same unit price as Brand E? (Hint: Use the rounded value you found in part a.)

5. To finish your grocery trip, you have to make choices with your remaining money. What is one possible combination of items you could purchase assuming you have $20.00 left to spend, you must purchase at least one of each item listed, and your total must be greater than $18.00. Justify your answer by showing your total spent and explain why you chose the items you did.

    One pound of bananas for $0.64
    A 5-lb bag of apples for $5.63
    A 3-lb bag of potatoes for $3.68
    A 2-lb bag of baby carrots for $1.99
    A container of guacamole for $2.67

CHAPTER 4

Ratios, Proportions, and Percents

4.1 Ratios and Unit Rates

4.2 Proportions

4.3 Decimals and Percents

4.4 Fractions and Percents

4.5 Solving Percent Problems Using Proportions

4.6 Solving Percent Problems Using Equations

4.7 Applications of Percent

4.8 Simple and Compound Interest

CHAPTER 4 PROJECTS

Take Me Out to the Ball Game!
How Much Will This Cell Phone Cost?

Connections

The concepts of ratios, proportions, and percents are used often in everyday life. Unit rates, a special type of ratio, are used to give the price per ounce of products in a grocery store so you can decide which is the better buy between two options. Proportions are used in cooking and baking to get the correct balance of ingredients in a recipe. And percents are used from the amount of active ingredient in a health product to the amount of interest you pay per month on your credit card. No matter your career or stage in life, ratios, proportions, and percents play an active role in your day-to-day activities. Homeowners often need to purchase home-improvement supplies such as paint, fertilizer, and grass seed for repairs and maintenance. In many cases, a homeowner must purchase supplies in quantities greater than what is needed for a particular job because the manufacturer's or distributor's packaging sizes do not fit the homeowner's exact needs. When this happens, the store will typically not sell part of a can of paint or a part of a bag of fertilizer. Fortunately, a little application of mathematics can help stretch any homeowner's dollar by minimizing any excess amounts that must be bought.

Suppose that during the late spring, you decide to treat your lawn with a fertilizer and weed killer combination. One bag contains 18 pounds with a recommended coverage of 5000 square feet. If your lawn is in the shape of a rectangle that is 150 feet long by 220 feet wide, how many pounds of the fertilizer and weed killer combination do you need to cover the lawn? How many bags will you need to purchase?

4.1 Ratios and Unit Rates

Ratios

A **ratio** is a comparison of _____
The ratio of *a* to *b* can be written as

_____

DEFINITION

Ratios have the following characteristics.

1. Ratios can be _____

2. The common units in a ratio can be _____

3. Generally, ratios are written with whole numbers in the _____

_____

A **rate** is a ratio with different _____

Changing Rates to Unit Rates

To make a rate a unit rate, _____

PROCEDURE

▶ Watch and Work

Watch the video for Example 9 in the software and follow along in the space provided.

Example 9 Application: Writing a Unit Rate

A bicyclist rides over level ground at a steady speed of 27 miles (mi) in 2.25 hours (hr). Find her speed in miles per hour.

Solution

✏️ Now You Try It!

Use the space provided to work out the solution to the next example.

Example A Application: Writing a Unit Rate

A swimmer swims laps at a steady pace of 6 laps in $\frac{1}{4}$ hour. Find his speed in laps per hour.

4.1 Exercises

Concept Check

True/False. Determine whether each statement is true or false. If a statement is false, explain how it can be changed so the statement will be true. (**Note:** There may be more than one acceptable change.)

1. The units in the numerator and denominator of a ratio must be the same, or need to be able to be converted to the same units.

2. The order of the numbers in a ratio or a rate is irrelevant as long as the numbers are reduced.

3. The ratio 8:2 can be reduced to the ratio 4.

4. To make a unit rate, divide the numerator by the denominator.

Practice

5. Write 18 to 28 as a fraction in lowest terms.

6. Write 5 nickels to 3 quarters as a fraction in lowest terms by first finding common units in the numerator and denominator.

7. Write the rate $200 in profit to $500 invested as a fraction in lowest terms.

8. Write the rate 270 miles to 4.5 hours as a unit rate.

Find the unit price (to the nearest tenth of a cent) of the set of items and tell which one is the better (or best) purchase.

9. *Coffee beans:* 1.75 oz at $1.99, 12 oz at $7.99

Applications

Solve.

10. A serving of three home-baked chocolate chip cookies weighs 40 grams and contains 12 grams of fat. What is the ratio, in lowest terms, of fat grams to total grams?

11. In recent years, 18 out of every 100 students taking the SAT (Scholastic Aptitude Test) at a local school have scored 600 or above on the mathematics portion of the test. Write the ratio, in lowest terms, of the number of scores 600 or above to the number of scores below 600.

Writing & Thinking

12. Demonstrate three different ways the ratio comparing 5 apples to 3 apples can be written. Choose one form and explain why it is the preferred form when using ratios in a math course.

13. When finding price per unit, will monetary units be located in the numerator or the denominator of the rate?

Name: _____ Date: _____ **119**

4.2 Proportions

Proportions

A **proportion** is a statement that _____

In symbols,
_____

A proportion is true if _____

DEFINITION

Solving a Proportion

1. Find the cross products (or cross multiply) and then _____

2. Divide both sides of the equation by _____

3. _____

PROCEDURE

▶ Watch and Work

Watch the video for Example 3 in the software and follow along in the space provided.

Example 3 Solving Proportions

Find the value of x if $\dfrac{4}{8} = \dfrac{5}{x}$.

Solution

✏️ Now You Try It!

Use the space provided to work out the solution to the next example.

Example A Solving Proportions

Find the value of x if $\dfrac{12}{x} = \dfrac{9}{15}$.

Solving an Application Using a Proportion

1. Identify the unknown quantity and _____

2. Set up a proportion in which the _____. (Make sure that the _____
   _____)

3. Solve the _____

PROCEDURE

4.2 Exercises

Concept Check

True/False. Determine whether each statement is true or false. If a statement is false, explain how it can be changed so the statement will be true. (**Note:** There may be more than one acceptable change.)

1. A proportion is a statement that two ratios are being multiplied.

2. Cross canceling is used to determine if a proportion is true.

3. In order to solve the proportion $\dfrac{16}{36.8} = \dfrac{x}{27.6}$ we construct the equation $36.8x = 441.6$.

4. When using proportions to solve a word problem, there is only one correct way to set up the proportion.

5. The proportions $\dfrac{36 \text{ tickets}}{\$540} = \dfrac{x \text{ tickets}}{\$75}$ and $\dfrac{x \text{ tickets}}{36 \text{ tickets}} = \dfrac{\$75}{\$540}$ will yield the same answer.

Practice

Determine whether each proportion is true or false.

6. $\dfrac{3}{6} = \dfrac{4}{8}$

7. $\dfrac{1}{3} = \dfrac{33}{100}$

Solve each proportion.

8. $\dfrac{5}{4} = \dfrac{x}{8}$

9. $\dfrac{3.5}{2.6} = \dfrac{10.5}{B}$

Applications

Solve.

10. The quality of concrete is based on the ratio of bags of cement to cubic yards of gravel. One batch of concrete consists of 27 bags of cement mixed into 9 cubic yards of gravel, while a second has 15 bags of cement mixed with 5 cubic yards of gravel. Determine whether the ratio of cement to gravel is the same for both batches.

11. An English teacher must read and grade 27 essays. If the teacher takes 20 minutes to read and grade 3 essays, how much time will he need to grade all 27 essays?

Writing & Thinking

12. In your own words, clarify how you can know that a proportion is set up correctly or not.

4.3 Decimals and Percents

The word percent comes from the Latin *per centum*, meaning _____. So **percent** means _____, or **the ratio of a number to** _____.

The symbol % is called the _____. This sign has the same meaning as the fraction $\frac{1}{100}$.

Changing a Decimal Number to a Percent

1. Move the _____

2. Write the _____

PROCEDURE

Changing a Percent to a Decimal Number

1. Move the _____

2. Delete the _____

PROCEDURE

▶ Watch and Work

Watch the video for Example 3 in the software and follow along in the space provided.

Example 3 Changing Percents to Decimal Numbers

Change each percent to a decimal number.

a. 76.%

b. 18.5%

c. 50%

d. 100%

e. 0.25%

✏️ Now You Try It!

Use the space provided to work out the solution to the next example.

Example A Changing Percents to Decimal Numbers

Change each percent to a decimal number.

a. 40%

b. 211%

c. 0.6%

d. 29.37%

e. 102%

4.3 Exercises

Concept Check

True/False. Determine whether each statement is true or false. If a statement is false, explain how it can be changed so the statement will be true. (**Note:** There may be more than one acceptable change.)

1. If a decimal number is less than 1, then the equivalent percent will be less than 100%.

2. It is not possible to have a percent greater than 100%.

3. A decimal number that is between 0.01 and 0.10 is between 10% and 100%.

4. To change from a percent to a decimal, simply omit the percent sign.

Practice

Change each fraction to a percent.

5. $\dfrac{20}{100}$

6. $\dfrac{125}{100}$

Change each decimal number to a percent.

7. 0.02

8. 2.3

4.3 Exercises

Change each percent to a decimal number.

9. 7%

10. 179%

Applications

Solve.

11. A savings account is offering an interest rate of 0.04 for the first year after opening the account. Change 0.04 to a percent.

12. Suppose that sales tax is figured at 7.25%. Change 7.25% to a decimal.

Writing & Thinking

13. Describe the relationship between percent and the number 100.

14. Describe a situation where more than 100% is possible. Describe a situation where it is impossible to have more than 100%.

4.4 Fractions and Percents

> **Changing a Fraction to a Percent**
> 1. Change the _____. (Divide _____.)
> 2. Change the _____.
>
> PROCEDURE

▶ Watch and Work

Watch the video for Example 3 in the software and follow along in the space provided.

Example 3 Changing Mixed Numbers to Percents

Change $2\frac{1}{4}$ to a percent.

Solution

✏ Now You Try It!

Use the space provided to work out the solution to the next example.

Example A Changing Mixed Numbers to Percents

Change $1\frac{1}{2}$ to a percent.

Changing a Percent to a Fraction or a Mixed Number

1. Write the percent as a fraction with _____

2. Reduce the _____

PROCEDURE

4.4 Exercises

Concept Check

True/False. Determine whether each statement is true or false. If a statement is false, explain how it can be changed so the statement will be true. (**Note:** There may be more than one acceptable change.)

1. Fractions that have denominators other than 100 cannot be changed to a percent.

2. The fraction $\frac{1}{5}$ is equivalent to $\frac{1}{5}\%$.

3. When changing from a percent to a mixed number, the fraction does not need to be reduced.

4. $75\% = 0.75 = \frac{3}{4}$

Practice

Change each fraction or mixed number to a percent. If necessary, round to the nearest tenth of a percent.

5. $\dfrac{3}{4}$

6. $5\dfrac{3}{10}$

Change each percent to a fraction or mixed number and reduce, if possible.

7. 120%

8. 12.5%

Applications

Solve.

9. Out of a possible total of 240 points on an exam, David received 204 points. What percent of the exam did David get correct?

10. To receive a Bachelor of Science (BS) degree at a certain college, the student must complete a total of 128 credit hours, of which 41 of these credits must be general education Core Skills courses. What percent of the total curriculum is dedicated to general education courses?

Writing & Thinking

11. Justify why mixed numbers are a larger percentage than proper fractions alone. (Consider the value of 100%.)

12. Describe the process to change a percent to a fraction or mixed number.

4.5 Solving Percent Problems Using Proportions

The Percent Proportion $\frac{P}{100} = \frac{A}{B}$

For the proportion $\frac{P}{100} = \frac{A}{B}$

P% = _____ (written _____).

B = _____ (number that _____).

A = _____ (a part of _____).

FORMULA

Three Basic Types of Percent Problems and the Proportion $\frac{P}{100} = \frac{A}{B}$

Type 1: Find the amount given the _____

What is _____

For example, what is _____

Type 2: Find the base given the _____

P% of what _____

For example, _____

Type 3: Find the percent given the _____

What percent of _____

For example, what _____

FORMULA

▶ Watch and Work

Watch the video for Example 1 in the software and follow along in the space provided.

Example 1 Finding the Amount

What is 65% of 500?

Solution

✏ Now You Try It!

Use the space provided to work out the solution to the next example.

Example A Finding the Amount

What is 15% of 80?

4.5 Exercises

Concept Check

True/False. Determine whether each statement is true or false. If a statement is false, explain how it can be changed so the statement will be true. (**Note:** There may be more than one acceptable change.)

1. Percent problems can be solved with a proportion if two of the three parts P, A, and B are known.

2. In the proportion $\dfrac{P}{100} = \dfrac{65}{200}$, the base is 65.

3. In the problem "What is 26% of 720?" the missing number is the base.

4. Because the base represents the whole, it is always larger than the amount.

Practice

Use the proportion $\dfrac{P}{100} = \dfrac{A}{B}$ to find each unknown quantity. Round percents to the nearest tenth of a percent. All other answers should be rounded to the nearest hundredth, if necessary.

5. Find 15% of 50.

6. What is 85% of 60?

7. 25% of 60 is _____.

8. What percent of 48 is 12?

9. _____% of 56 is 140.

Applications

Solve.

10. In 2016, the Los Angeles Dodgers led the major leagues in home attendance, drawing an average of 45,720 fans to their home games. This figure represented 81.64% of the capacity of Dodger Stadium. Estimate how many fans the stadium can hold (to the nearest ten) when it is filled to capacity.[1]

11. You want to purchase a new home for $122,000. The bank will loan you 80% of the purchase price. How much will the bank loan you? (This amount is called your mortgage and you will pay it off over several years with interest. For example, a 30-year loan will probably cost you a total of more than 3 times the original loan amount.)

Writing & Thinking

12. List the four parts of the proportion equation and give a brief definition of each one.

13. Can a mixed number be used in a proportion? Justify your answer.

---

[1] Source: espn.go.com/mlb/attendance

4.6 Solving Percent Problems Using Equations

Terms Related to the Basic Equation $R \cdot B = A$

For the basic equation $R \cdot B = A$,

$R = $ _____ (as a _____).

$B = $ _____ (number that we _____).

$A = $ _____ (a part of _____).

DEFINITION

Three Basic Types of Percent Problems and the Formula $R \cdot B = A$

Type 1: Find the amount given the _____

For example, what is _____

_____

Type 2: Find the base given the _____

For example, 42% of what _____

_____

Type 3: Find the percent (rate) given the _____

For example, what _____

_____

FORMULA

▶ Watch and Work

Watch the video for Example 4 in the software and follow along in the space provided.

Example 4 Finding the Amount

Find 75% of 56.

Solution

✏ Now You Try It!

Use the space provided to work out the solution to the next example.

Example A Finding the Amount

Find 150% of 60.

4.6 Exercises

Concept Check

True/False. Determine whether each statement is true or false. If a statement is false, explain how it can be changed so the statement will be true. (**Note:** There may be more than one acceptable change.)

1. In order to solve the equation $0.56 \cdot B = 12$ for the base B one would multiply 12 by 0.56.

2. In the problem "126% of 720 is what number?" the missing number is the amount.

3. The solution to the problem "50% of what number is 352?" could be found by solving the equation $50 \cdot B = 352$.

4. If the base is 120 and the rate is greater than 100%, then the amount will be greater than 120.

Practice

Use the equation $R \cdot B = A$ to find each unknown quantity. Round percents to the nearest tenth of a percent. All other answers should be rounded to the nearest hundredth, if necessary.

5. 10% of 70 is what number?

6. Find 75% of 12.

7. 150% of _____ is 63.

8. What percent of 75 is 15?

9. _____% of 30 is 6.

Applications

Solve.

10. During his presidency, from 1945 to 1953, Harry S. Truman vetoed 250 congressional bills, and 12 of those vetoes were overridden. What percent of Truman's vetoes were overridden?

11. The minimum down payment to obtain the best financing rate on a house is 20%. Assuming that John has set aside $35,000 and wants to take advantage of the best financing rate, what is the most expensive house he can purchase?

Writing & Thinking

12. Explain the connection between the proportion $\frac{P}{100} = \frac{A}{B}$ and the equation $R \cdot B = A$.

13. Explain how to determine which number is the rate (percent), which one is the amount, and which one is the base.

4.7 Applications of Percent

Basic Steps for Solving Word Problems
1. _____
2. _____
3. _____
4. _____

PROCEDURE

Terms Related to Discount

Discount: difference between the _____

Sale price: original price minus _____

Rate of discount percent of _____

DEFINITION

Terms Related to Sales Tax

Sales tax: _____

Rate of sales tax: _____

DEFINITION

▶ Watch and Work

Watch the video for Example 3 in the software and follow along in the space provided.

Example 3 Application: Solving Sales Tax Problems

If the rate of sales tax is 6%, what would be the final cost of a laptop priced at $899?

Solution

4.7 Applications of Percent

✏️ Now You Try It!

Use the space provided to work out the solution to the next example.

Example A Application: Solving Sales Tax Problems

Assuming a 7% sales tax rate, what would be the final cost of a discounted pair of shoes priced at $39?

1. A **commission** is a _____.

2. At times, it is helpful to know by what percent the value changed. This is called finding the _____

   (or the _____).

Profit and Percent of Profit

Profit: the difference between _____

$$\text{Profit} = \underline{\hspace{3cm}}$$

Percent of Profit: There are two types; both are _____
1. Percent of profit **based on** _____

$$\frac{\text{Profit}}{\text{Cost}} = \% \text{ of profit based on } \underline{\hspace{2cm}}$$

2. Percent of profit **based on** _____

$$\frac{\text{Profit}}{\text{Selling Price}} = \% \text{ of profit based on } \underline{\hspace{2cm}}$$

FORMULA

4.7 Exercises

Concept Check

True/False. Determine whether each statement is true or false. If a statement is false, explain how it can be changed so the statement will be true. (**Note:** There may be more than one acceptable change.)

1. If an item is selling for a 35% discount, the customer will pay 65% of the original price.

2. If you must pay 7% sales tax on a purchase, the total cost you will pay is 170% of the total before tax.

3. A car was purchased in 1965 for $3800. It sold for $1200 in 2011. This is an example of depreciation.

4. Profit is determined by subtracting selling price from the cost.

Applications

Solve.

5. A new briefcase was priced at $275. If it were to be marked down 30%:

   a. What would be the amount of the discount?

   b. What would be the new price?

6. If sales tax is figured at 7.25%, how much tax will be added to the total purchase price of three textbooks priced at $25.00, $35.00, and $52.00?

7. A realtor works on 6% commission. What is his commission on a house he sold for $195,000?

8. The cost of an 85" smart TV to a store owner was $3300 and he sold the TV for $4500.

   a. What was his profit?

   b. What was his percent of profit based on cost?

   c. What as his percent of profit based on selling price?

Writing & Thinking

9. Determine how to calculate sales tax when eating out and relate this process to either a proportion and/or using the amount/base/rate equation. Give an example.

Name: Date:

4.8 Simple and Compound Interest

> **Simple Interest Formula**
>
> Interest = Principal · rate · time
>
> Writing the formula using letters, we have $I = P \cdot r \cdot t$, where
>
> $I =$ _____
>
> $P =$ _____
>
> $r =$ _____ in decimal or fraction form, and
>
> $t =$ _____
>
> **FORMULA**

▶ Watch and Work

Watch the video for Example 1 in the software and follow along in the space provided.

Example 1 Application: Calculating Simple Interest

You want to borrow $2000 from your bank for one year. If the interest rate is 5.5%, how much interest would you pay?

Solution

✏ Now You Try It!

Use the space provided to work out the solution to the next example.

Example A Application: Calculating Simple Interest

If you were to borrow $1500 at 8.5% for one year, how much interest would you pay?

144   4.8   Simple and Compound Interest

Calculating Compound Interest

1. Use the formula _____

   Let $t = \dfrac{1}{n}$ where n _____
2. Add this interest to the _____
3. Repeat steps _____

PROCEDURE

Compound Interest Formula

When interest is compounded, the total **amount** A accumulated (including principal and interest) is given by the formula

$$A = P\left(1 + \frac{r}{n}\right)^{nt},$$

where

$P = $ _____

$r = $ _____

$t = $ _____

$n = $ _____

FORMULA

Total Interest Earned

To find the total interest earned on an investment that has earned interest by compounding, _____
_____

$I = $ _____

FORMULA

Inflation

The adjusted amount A due to **inflation** is

$$A = P(1 + r)^t,$$

where

$P = $ _____

$r = $ _____

$t = $ _____

FORMULA

> **Depreciation**
>
> The current value V of an item due to **depreciation** is
>
> $$V = P(1-r)^t,$$
>
> where
>
> $P = $ _____
> $r = $ _____
> $t = $ _____
>
> FORMULA

4.8 Exercises

Concept Check

True/False. Determine whether each statement is true or false. If a statement is false, explain how it can be changed so the statement will be true. (**Note:** There may be more than one acceptable change.)

1. In the simple interest formula, the rate can be written as a decimal number or a fraction.

2. Simple interest can be compounded monthly or quarterly.

3. Interest cannot be earned on interest, only the principal.

4. Inflation can be treated in the same manner as simple interest.

Applications

Solve. Round to the nearest cent, if necessary.

5. How much interest would be paid on a loan of $3000 at 5% for 9 months?

4.8 Exercises

Solve each problem by repeatedly using the formula for calculating simple interest. Round your answer to the nearest cent, if necessary.

6. You loan your cousin $2000 at 5% compounded annually for 3 years. How much interest will your cousin owe you?

   a. First year: $I = 2000 \cdot 0.05 \cdot 1 =$ _____

   b. Second year: $I =$ _____ $\cdot\, 0.05 \cdot 1 =$ _____

   c. Third year: $I =$ _____ $\cdot\, 0.05 \cdot 1 =$ _____

   d. The total interest is _____ .

Solve each problem by using the compound interest formula. Round your answer to the nearest cent, if necessary.

7. You deposit $1500 at 4% to be compounded semiannually. How much interest will you earn in 3 years?

Solve. Round your answer to the nearest cent, if necessary.

8. Gavin bought a new truck last year for $29,900. This year, he decided that he wants to trade it in for a smaller car. He can resell the truck for 26,500. What was the rate of depreciation for the year?

Writing & Thinking

9. List the four parts involved in the simple interest formula. In your own words, define each one.

10. Compare and contrast simple interest with compound interest.

Chapter 4 Project

Take Me Out to the Ball Game!

An activity to demonstrate the use of percents and percent increase or decrease in real life.

The Atlanta Braves baseball team has been one of the most popular baseball teams for fans, not only from Georgia, but throughout the Carolinas and the southeastern United States. The Braves franchise started playing at the Atlanta-Fulton County Stadium in 1966 and this continued to be their home field for 30 years. In 1996, the Centennial Olympic Stadium that was built for the 1996 Summer Olympics was converted to a new ballpark for the Atlanta Braves. The ballpark was named Turner Field and was opened for play in 1997. In 2017, the Braves moved to a new stadium named SunTrust Park, which was renamed as Truist Park in 2020 when SunTrust and BB&T merged.

Round all percents to the nearest whole percent.

1. The Atlanta-Fulton County Stadium had a seating capacity of 52,769 fans. Turner Field had a seating capacity of 50,096. Truist Park has a seating capacity of 41,149.
   a. Determine the decrease in seating capacity between Turner Field and the original Braves stadium.
   b. Determine the percent decrease in seating capacity between Truist Park and Turner field.

2. The Centennial Olympic Stadium had approximately 85,000 seats. Some of the seating was removed in order to convert it to the Turner Field ballpark. Rounding the number of seats in Turner Field to the nearest thousand, what is the approximate percent decrease in seating capacity from the original Olympic stadium?

3. The Braves were ranked 2nd out of 16 teams in 1999 and the average attendance was 40,554. Again in 2021, they were ranked 2nd out of 15 teams and the average attendance was 28,746. What was the percent decrease in attendance from 1999 to 2021? [1]

4. The highest average attendance for the Braves was 47,960 in 1993 at the Atlanta-Fulton County Stadium. The lowest average attendance was 6642 in 1975 at the Atlanta-Fulton County Stadium. What is the percent increase from the lowest attendance to the highest? [2]

5. Chipper Jones, a popular Braves third baseman, retired in July 2013. He started his career with the Braves in 1993 at the age of 21. [3]
   a. In 2001, Chipper had 189 hits in 572 at-bats. Calculate Chipper's batting average for the season by dividing the number of hits by the number of at-bats. Round to the nearest thousandth.
   b. In 2008, Chipper had 160 hits in 439 at-bats. Calculate Chipper's batting average for the season by dividing the number of hits by the number of at-bats. Round to the nearest thousandth.
   c. Calculate the percent change in Chipper's batting average from 2001 to 2008.
   d. Does this represent a percent increase or decrease?

6. In 2001, Chipper had 102 RBIs (runs batted in). In 2008, Chipper had only 75 RBIs.
   a. Calculate the percent change in RBIs from 2001 to 2008.
   b. Does this represent a percent increase or decrease?

---

[1] Source: "Atlanta Braves Attendance, Stadiums, and Park Factors," Baseball Reference, Accessed May 3, 2022, www.baseball-reference.com/teams/ATL/attend.shtml.
[2] Source: "Atlanta Braves Attendance Data," Baseball Almanac, Accessed May 3, 2022, baseball-almanac.com/teams/bravatte.shtml.
[3] Source: "Chipper Jones Stats," Baseball Almanac, Accessed May 3, 2022, www.baseball-almanac.com/players/player.php?p=jonesch06.

Chapter 4 Project

How Much Will This Cell Phone Cost?

An activity to demonstrate the use of percent increases and decreases in real life.

Sergio works in electronics sales. He wants to purchase a new cell phone from his place of employment, which offers a 10% employee discount. Sales tax is 7% in the city where the store is located. The original price of the cell phone Sergio wants to buy is $500.

For the following problems, round monetary amounts to the nearest cent. Otherwise, do not round unless indicated.

1. Suppose the store calculates the discount before applying sales tax.
   a. What is the price of the phone following the 10% discount, before tax?
   b. What is the total after applying tax to the answer in part a.?

2. Let's compare the price Sergio pays to the original price of the phone.
   a. Does the price Sergio pays represent an increase or a decrease from the original price? By how much?
   b. What percent of the original price for the phone is the price that Sergio pays in Problem 1 part a.?
   c. Does the answer to part b. represent a percent increase or a percent decrease from the original price?
   d. Calculate the percent increase or percent decrease of the price of the phone.

3. Consider whether the order of applying the discount and applying the tax matters.
   a. Find the cost of the phone if the tax is applied first, followed by the discount.
   b. How does the cost found in part a. compare to the cost found in Problem 1 part b.?

4. Sergio's coworker Haruto is looking at a cell phone with an original price of $450. Consider whether this change in original price will affect the *percent* of the original price he will pay after the employee discount.
   a. Repeat Problem 1 parts a. and b. and Problem 2 part b. for the phone Haruto is interested in.
   b. Does the value from part a. represent a percent increase or a percent decrease from the original price? By how much?
   c. How does the value in part b. compare to the value from Problem 2 part d.?

5. Zendaya works at the same store as Sergio and Haruto. The final price she pays for a cell phone, including both the discount and tax, is $311.53. What was the original price for her cell phone?

6. Marya works at another electronics store in the same city (which means the sales tax is also 7%), but her employer only offers a 5% employee discount. Like Sergio, she is looking at a cell phone with an original price of $500.
   a. Repeat Problem 1 parts a. and b. and Problem 2 part b. for the phone Marya is interested in.
   b. Does this represent a percent increase or a percent decrease? By how much?
   c. How do the percents for the employee discount and tax determine if there is a percent increase or a percent decrease?

7. Suppose Sergio does not have the money on hand to pay the value calculated in Problem 1 part b. in full, so he decides to finance the phone on a 2-year plan that will charge a 30% interest rate, compounded monthly.

   a. Determine the final cost of the phone when Sergio has paid it off in full after 2 years. Round to the nearest hundredth. (**Hint:** You will need to use the compound interest formula.)

   b. What percent of the original price of $500 will Sergio have paid?

   c. Does this represent a percent decrease or a percent increase? By how much?

8. What if Sergio trades in his previous cell phone for $100? Calculate the final cost of the new phone after paying it off using the same 2-year financing plan and determine by how much there is a percent increase or decrease when compared to the original price of $500. Round to the nearest hundredth of a percent.

CHAPTER 5

Measurement

5.1 US Measurements

5.2 The Metric System: Length and Area

5.3 The Metric System: Capacity and Weight

5.4 US and Metric Equivalents

CHAPTER 5 PROJECTS
Metric Cooking
Confused Conversions

Connections

Measurements surround us as we go about our daily lives. When you drive to school, work, or a friend's house, you must obey the speed limits that are posted in miles per hour. When you fill your car with gasoline, the fuel pump keeps track of how many gallons it dispenses. When you cook, the ingredients are measured by weight or volume. All of these measurements play a role in how you interact with the world around you. Knowing how different units of measurement work together will help you be more successful in your day-to-day life.

Suppose you purchase a French cookbook to expand your culinary skills. The book you purchased has recipes that are a direct translation from French, meaning that the ingredient measurements and cooking temperatures are given in metric units. For example, the recipe for *Pain au Chocolat* calls for 120 milliliters of milk and needs the oven to be preheated to 190 °C. If you are living in the United States, you might only have measuring cups that measure fluid ounces and an oven that has a temperature dial in degrees Fahrenheit. How can you determine how many cups of milk you need for the recipe and to what temperature to preheat your oven?

5.1 US Measurements

Using Multiplication and Division to Convert Measurements
1. Multiply to convert to _____. (There will be _____.)
2. Divide to convert to _____. (There will be _____.)

PROCEDURE

Using Unit Fractions to Convert Measurements
1. The numerator should be in the _____
2. The denominator should be in the _____

PROCEDURE

▶ Watch and Work

Watch the video for Example 6 in the software and follow along in the space provided.

Example 6 Application: Converting US Units of Measure

Determine how many seconds are in a 5-day work week assuming an 8 hr work day.

Solution

✏ Now You Try It!

Use the space provided to work out the solution to the next example.

Example A Application: Converting US Units of Measure

How many fluid ounces are in 8 gallons of apple juice?

5.1 Exercises

Concept Check

True/False. Determine whether each statement is true or false. If a statement is false, explain how it can be changed so the statement will be true. (**Note:** There may be more than one acceptable change.)

1. Capacity can be measured using ounces, quarts, and gallons.

2. One mile is equivalent to 2000 feet.

3. To convert from smaller units to larger units, division will be required.

4. Multiplication by a unit fraction does not change the value of the expressions being converted.

Practice

Convert each measurement.

5. 4 pt = ___ c

6. 10 mi = ___ ft

7. 39 ft = ___ yd

8. 150 min = ___ hr

Applications

Solve.

9. Sheer fabric costs $7.99 per yard. If it will take 35 feet of fabric to make drapes for the entire house, how much must you spend on fabric for the drapes, to the nearest cent?

10. The author of this textbook spent 1 year, 23 weeks, 5 days, and 14 hours writing it. How many seconds is this? (**Hint:** There are 52 weeks in a year.)

Writing & Thinking

11. Colby needs to find out how many yards are in one mile. What two sets of equivalent units would he need to make that determination?

12. In your own words, explain when you would multiply and when you would divide when converting between units.

Name: _____ Date: _____ **155**

5.2 The Metric System: Length and Area

Writing Metric Units of Measure

In the metric system,

1. A 0 is written to the left of the decimal point if _____

   For example, _____

2. No commas are used in writing numbers. If a number has more than four digits (to the left or right of the decimal point), the digits are _____

   For example, _____

PROCEDURE

There are two basic methods of converting units of measurement in the metric system:

1. multiplying by_____,
2. moving the _____

Using Unit Fractions to Convert Measurements

1. The numerator should be in the _____
2. The denominator should be in the _____

PROCEDURE

▶ Watch and Work

Watch the video for Example 6 in the software and follow along in the space provided.

Example 6 Converting Metric Units of Area

Convert each measurement using unit fractions.

a. $5 \text{ cm}^2 =$ _____ mm^2

b. $4600 \text{ mm}^2 =$ _____ m^2

Solution

Now You Try It!

Use the space provided to work out the solution to the next example.

Example A Converting Metric Units of Area

Convert each measurement using unit fractions.

a. $86 \text{ m}^2 = $ _____ cm^2

b. $0.06 \text{ mm}^2 = $ _____ dm^2

5.2 Exercises

Concept Check

True/False. Determine whether each statement is true or false. If a statement is false, explain how it can be changed so the statement will be true. (**Note:** There may be more than one acceptable change.)

1. To change from smaller units to larger units, multiplication must be used.

2. Units of length in the metric system are named by putting a prefix in front of the basic unit meter, for example, centimeter.

3. In metric units, a square that is 1 centimeter long on each side is said to have an area of 1 centimeter.

Practice

Convert each measurement.

4. 3 m = ___ cm

5. 19.77 m = ___ km

6. 6 500 000 hertz = ___ megahertz

7. 13 dm² = _____ cm² = _____ mm²

Applications

Solve.

8. A triangle has a base measuring 4 cm and a height measuring 16 mm. Determine the area of the triangle in cm².

9. A section of railroad track measuring 2.1 km in length needs to be replaced. Each railroad tie is 4 decimeters wide and they are to be spaced 0.8 m apart. How many railroad ties will be needed to complete this section of track?

Writing & Thinking

10. Compare and contrast ease of converting units in the US customary system and the metric system.

11. Discuss the meaning of prefixes like milli-, centi-, and kilo- in metric units. Give examples.

5.3 The Metric System: Capacity and Weight

1. In the metric system, capacity (liquid volume) is measured in _____ (abbreviated _____).
2. A liter is the volume enclosed in a cube that is _____ on each edge.
3. Mass is _____ in an object.
4. The basic unit of mass in the metric system is the _____.

▶ Watch and Work

Watch the video for Example 7 in the software and follow along in the space provided.

Example 7 Converting Metric Units of Weight

Convert 34 g to milligrams **a.** using a unit fraction and **b.** using a metric conversion line.

Solution

Now You Try It!

Use the space provided to work out the solution to the next example.

Example A Converting Metric Units of Weight

Convert 14.9 kg to grams using a unit fraction or a metric conversion line.

5.3 Exercises

Concept Check

True/False. Determine whether each statement is true or false. If a statement is false, explain how it can be changed so the statement will be true. (**Note:** There may be more than one acceptable change.)

1. One milliliter is equivalent to one cubic centimeter.

2. Volume is measured in square units.

3. In 1 liter there are 100 milliliters.

4. A metric ton and a US customary ton are equal (a metric ton weighs about 2000 US pounds).

5. A dekagram contains 10 grams.

Practice

Convert each measurement.

6. 2 L = ___ mL

7. 6.3 kL = ___ L

8. 2 g = ___ mg

9. 2000 g = ___ kg

Applications

Solve.

10. How many 5-mL doses of liquid medication can be given from a vial containing 3 deciliters?

11. One cup of flour is approximately 120 grams. How many cups of flour can you get out of a bag of flour weighing 2.4 kg?

Writing & Thinking

12. In the metric system, the common unit of capacity is the liter. Discuss how you would change from a measure of liters to milliliters.

Name: _____ Date: _____ **163**

5.4 US and Metric Equivalents

Temperature

US customary measure is in _____

Metric measure is in _____

DEFINITION

Temperature Formulas

F = Fahrenheit temperature and C = Celsius temperature

$F = $ _____ $C = $ _____

FORMULA

▶ Watch and Work

Watch the video for Example 5 in the software and follow along in the space provided.

Example 5 Converting Units of Area

Convert each measurement, rounding to the nearest hundredth.

a. 40 yd² = _____ m²

b. 100 cm² = _____ in.²

c. 6 acres = _____ ha

d. 5 ha = _____ acres

Solution

✏️ Now You Try It!

Use the space provided to work out the solution to the next example.

Example A Converting Units of Area

Convert each measurement, rounding to the nearest hundredth.

a. 53 in.² = _____ cm²

b. 50 m² = _____ ft²

c. 16 acres = _____ ha

d. 3 ha = _____ acres

5.4 Exercises

Concept Check

True/False. Determine whether each statement is true or false. If a statement is false, explain how it can be changed so the statement will be true. (**Note:** There may be more than one acceptable change.)

1. Water freezes at 32 degrees Celsius.

2. When converting between US customary and metric units, often the results will be approximations.

3. A 5K (km) run is longer than a 5 mile run.

4. One square meter covers more area than one square yard.

Practice

Convert each measurement. Round to the nearest hundredth, if necessary.

5. 25 °C = ___ °F

6. 9 ft = ___ m

7. 3 in.² = _____ cm²

8. 4 qt = ___ L

9. 33 kg = ___ lb

Applications

Solve.

10. While visiting her aunt in Germany, Helga wants to surprise her aunt with a cake. She brought her mom's cake recipe with her from Georgia. The recipe says to bake the cake at 350 degrees Fahrenheit but the temperature gauge on her aunt's oven is in degrees Celsius. To what temperature should Helga set her aunt's oven in order to bake the cake at the correct temperature? Round the temperature to the nearest degree.

11. The Ironman Triathlon championship in Hawaii consists of a swim of 3.86 km, a bike ride of 180.25 km, and finishes with a run equal to the length of a standard marathon. A marathon is typically 26.2 miles. What is the total length of the Ironman Triathlon in kilometers? Round the length to the nearest tenth of a km.

Writing & Thinking

12. Most conversions between the US customary system of measure and metric system are not exact. Explain why this is true and give any exceptions.

Name: _____   Date: _____   **167**

Chapter 5 Project

Metric Cooking
An activity to demonstrate the use of metric to US conversions in real life.

Your grandma lives outside of the United States and has emailed you her famous apple pie recipe to use at an upcoming party. As you start to bake the pie on the day of the party, you realize grandma's recipe is written using only metric units. Looking through the supplies in your kitchen, you find a scale that measures weight in ounces and some measuring spoons that measure volume in $\frac{1}{8}$, $\frac{1}{4}$, and $\frac{1}{2}$ of a teaspoon and 1 whole tablespoon. In order to successfully bake the pie in time for the party, you must quickly convert the metric measurements to US measurements. You start with the pie crust ingredients.

Pie Crust

Ingredient	Metric Measurement	US Measurement
Flour	280 g	_____ oz
Vegetable Shortening	90 g	_____ oz
Unsalted Butter	50 g	_____ oz
Cold Water	90 mL	_____ tbsp
Salt	5 mL	_____ tsp

Apple Filling

Ingredient	Metric Measurement	US Measurement
Apples	1 kg	_____ oz
Sugar	100 g	_____ oz
Cornstarch	15 mL	_____ tsp
Cinnamon	2.5 mL	_____ tsp
Salt	0.5 mL	_____ tsp
Nutmeg	0.5 mL	_____ tsp
Butter	25 g	_____ oz

1. Fill in the third column of the pie crust table by converting the measurements of each ingredient using the correct conversion factors. Use 1 mL = 0.068 tbsp and 1 mL = 0.203 tsp. (**Note:** tbsp stands for tablespoon and tsp stands for teaspoon.)

2. The recipe requires the oven to be preheated to 230 °C, but your oven measures degrees in Fahrenheit. What temperature should you preheat your oven to?

   While the oven is preheating, you begin to prepare the ingredients for the apple filling.

3. Fill in the third column of the apple filling table by converting the measurements of each ingredient using the correct conversion factors. Use 1 mL = 0.068 tbsp and 1 mL = 0.203 tsp.

4. Since the measuring spoons can only measure $\frac{1}{8}$, $\frac{1}{4}$, and $\frac{1}{2}$ of a teaspoon and 1 whole tablespoon, what is the most reasonable way to round each US volume measurement in each of the tables?

In addition to the usual recipe, your grandma has listed several variations on the recipe that depend on the taste of the apples. For bland apples, you should add 20 mL of lemon juice to the filling. For sour apples, you should increase the sugar to 140 g.

5. You taste the apples and decide you need to increase the sugar to 140 g. You have already added 100 g. How many more ounces of sugar do you need to add to reach 140 g?

6. You notice that the recipe requires a pie pan with a diameter of 23 cm. After measuring the pie pan you discover it is 9 inches in diameter. Is the pie pan the right size for the apple pie? What is its diameter in inches? Round to the nearest whole number.

Chapter 5 Project

Confused Conversions

An activity to investigate the proper use of conversions between systems of measurements in real life.

Imagine talking to your European friend and you remark that the daily high was 30 degrees. Things can get a little confusing: 30 °F is a cold snap while 30 °C is a very warm summer day. This confusion could be avoided by making sure that you are clear about which system's units you are using.

Could confusing measurements between the metric system and the US system cause more harm than just a misunderstanding? We will explore some real-life situations where things didn't go very well.

1. In 1983, Air Canada Flight 143, the infamous Gimli Glider made an emergency landing in Gimli, Manitoba, when it ran out of fuel midair. The mistake can be traced to a refueling procedure: the plane was supposed to be fueled using kilograms, but it was instead refueled using pounds.

   a. Considering that there are approximately 0.454 kilograms in one pound, determine the number of kilograms in 100 pounds of fuel.

   b. Suppose an empty tank can hold 400 kilograms of fuel. Instead, 400 pounds of fuel are added to the tank. How many more kilograms of fuel would be needed to get a full tank?

2. In 1998, NASA had a slight miscommunication with the Mars Climate Orbiter. This was a $125-million spacecraft designed to study Mars's atmosphere. It was forever lost in space once it was made to accelerate too quickly.

   c. The standard metric unit for impulse is the Newton-second (N · s) while the customary English unit is the pound-second (lb · s). We know that one Newton is approximately equal to 4.45 pounds. What is the value of 250 N · s in lb · s?

   d. The Mars Climate Orbiter was calibrated to receive impulse information in N · s but NASA inadvertently converted the number to lb · s. How many times larger was the impulse value that was relayed to the orbiter compared to the correct value.

3. A grain is a unit of measure equal to about 0.065 grams. In 1999, the Institute for Safe Medication Practices reported a case of a patient who received 0.5 *grams* of phenobarbital (a sedative) instead of the prescribed 0.5 *grains*.

   e. Determine the dose in grains of 0.5 grams of phenobarbital. Round to the nearest tenth.

   f. What is a possible consequence of giving a patient 0.5 grams instead of 0.5 grains?

4. Perform an internet search for "Verizon cents versus dollars." This is a video of a customer disputing their data charges with a cellphone carrier. Explain how his billing issue is related to our confused conversions problems.

5. Have you ever found yourself in a situation where a misunderstanding about units and conversions created a problem? Either describe your own situation or research one on the internet.

CHAPTER 6

Geometry

6.1 Angles and Triangles

6.2 Perimeter

6.3 Area

6.4 Circles

6.5 Volume and Surface Area

6.6 Similar and Congruent Triangles

6.7 Square Roots and the Pythagorean Theorem

CHAPTER 6 PROJECTS
Before and After

This Mixtape is Fire!

Connections

Sometimes geometry can feel a little removed from our day to day lives. You might catch yourself asking questions such as "When am I ever going to use the formula for the area of a rectangle?" or "Why do I need to know the characteristics of right triangles?" Upon closer inspection of our surroundings, it is very hard to deny that geometry is present in many aspects of modern life. This topic is present in a variety of situations, such as a family deciding how much paint to purchase to repaint their living room and a town government budgeting for the construction of a new school building.

The next time you walk through a building, look around and notice the geometric shapes that were used in its construction. The walls, ceiling, and floors are likely rectangles. Many stairways are supported by triangular structures. And windows can be squares, rectangles, or circles.

Suppose your college decides to paint all four walls of a classroom that is 25 ft by 20 ft with walls that are 10 ft high. The administration has selected a paint with a coverage of 350 ft^2 per gallon. How would you determine many gallons of paint will be needed to cover the classroom walls if you ignore all openings such as windows and doors?

6.1 Angles and Triangles

Point, Line, Plane

Undefined Term	Representation	Discussion
Point		A point is represented by _____
		Points are labeled with _____
Line	_____	A line has no _____
		Lines are labeled with _____ _____
Plane	_____	Flat surfaces, such as a table top or wall, represent _____ _____
		Planes are labeled with _____
	_____	

DEFINITION

Ray and Angle

Term	Definition	Illustrations with Notation
Ray	A ray consists of _____ _____ _____ _____	_____
Angle	An angle consists of _____ _____ _____ _____	_____

DEFINITION

174   6.1   Angles and Triangles

Labeling Angles

There are three common ways of labeling angles:

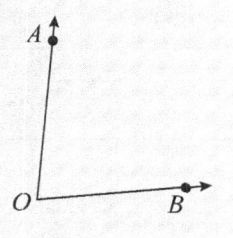

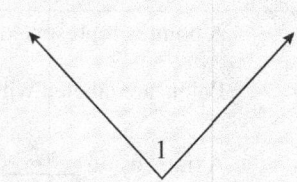

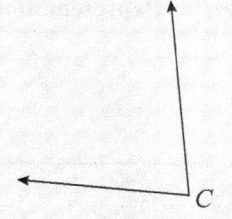

A. ∠_____   **B.** ∠___   **C.** ∠___

Using three _____   Using single _____   Using the single _____

PROCEDURE

1. The base unit when measuring angles is _____ (symbolized _____).

Angles Classified by Measure

Name	Measure	Illustrations with Notation
Acute	_____	
Right	_____	
Obtuse	_____	
Straight	_____	

DEFINITION

6.1 Angles and Triangles 175

Complementary and Supplementary Angles

1. Two angles are **complementary** if _____

2. Two angles are **supplementary** if _____

DEFINITION

2. If two angles have the same measure, they are said to be _____ (symbolized as ≅).

Vertical Angles

Vertical angles _____

That is, vertical angles have _____

DEFINITION

Adjacent Angles

Two angles are adjacent if _____
_____

DEFINITION

Parallel Lines and Perpendicular Lines

Term	Definition	Illustrations with Notation
Parallel Lines	Two lines are parallel (symbolized ∥) if _____ _____ _____ _____	$\overrightarrow{PQ}$ is parallel to _____ ($\overrightarrow{PQ}$ _____)
Perpendicular Lines	Two lines are perpendicular (symbolized ⊥) if _____ _____ _____	$\overrightarrow{PQ}$ is perpendicular to _____ ($\overrightarrow{PQ}$ _____)

DEFINITION

Parallel Lines and a Transversal

If two parallel lines are cut by a transversal, then the following two statements are true.

1. _____

2. _____

PROPERTIES

▶ Watch and Work

Watch the video for Example 8 in the software and follow along in the space provided.

Example 8 Calculating Measures of Angles

In the figure shown, lines k and l are parallel, t is a transversal, and $m\angle 1 = 50°$. Find the measures of the other 7 angles.

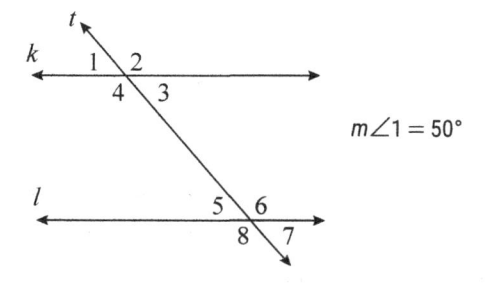

$m\angle 1 = 50°$

Solution

Now You Try It!

Use the space provided to work out the solution to the next example.

Example A Calculating Measures of Angles

In the figure shown, lines l and m are parallel, t is a transversal, and $m\angle 2 = 80°$. Find $m\angle 4$, $m\angle 5$, and $m\angle 6$.

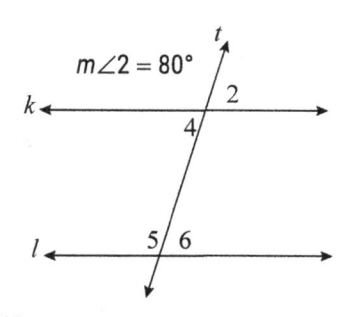

---

3. A **line segment** consists of _____.

4. A **triangle** consists of _____.

Triangles Classified by Sides

(Note: In the figures, sides with equal length are indicated by the same number of tic marks.)

Name	Property	Example
Scalene	_____ _____ _____	$\triangle ABC$ is scalene since _____ _____
Isosceles	_____ _____ _____	$\triangle PQR$ is isosceles since _____
Equilateral	_____ _____ _____	$\triangle XYZ$ is equilateral since _____

DEFINITION

178   6.1   Angles and Triangles

Triangles Classified by Angles

Name	Property	Example
Acute	_____ _____ _____	_____ △ABC is acute since _____
Right	_____ _____	△PRQ is a right triangle since _____
Obtuse	_____ _____ _____	△XYZ is an obtuse triangle since _____

DEFINITION

Three Properties of Triangles

In a triangle, the following statements are true:

1. The sum of the measures _____

2. The sum of the lengths of _____

3. Longer sides are _____

PROPERTIES

6.1 Exercises

Concept Check

True/False. Determine whether each statement is true or false. If a statement is false, explain how it can be changed so the statement will be true. (**Note:** There may be more than one acceptable change.)

1. The sum of the measures of two complementary angles is equal to the measure of one right angle.

2. The sum of the measures of complementary angles is greater than the sum of the measures of supplementary angles.

3. Adjacent angles are two angles that share a vertex and a common side but do not overlap.

4. If two lines in a plane are not parallel, then they are perpendicular.

5. A triangle with sides of 4 inches, 4 inches, and 3 inches is an isosceles triangle.

6. A triangle with three angles that each measure less than 90 degrees is an acute triangle.

Practice

7. Name the type of angle formed by the hands on a clock.

   a. at six o'clock

   b. at three o'clock

   c. at one o'clock

   d. at five o'clock

6.1 Exercises

8. Assume that ∠1 and ∠2 are complementary.
   a. If $m\angle 1 = 15°$, what is $m\angle 2$?
   b. If $m\angle 1 = 3°$, what is $m\angle 2$?
   c. If $m\angle 1 = 45°$, what is $m\angle 2$?
   d. If $m\angle 1 = 75°$, what is $m\angle 2$?

9. The figure shows two intersecting lines.
   a. If $m\angle 1 = 30°$, what is $m\angle 2$?

   b. Is $m\angle 3 = 30°$? Give a reason for your answer other than the fact that ∠1 and ∠3 are vertical angles.

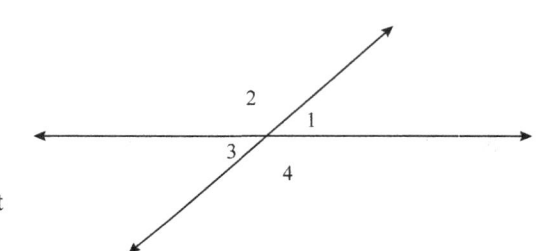

   c. Name two pairs of congruent angles.

   d. Name four pairs of adjacent angles.

6.1 Exercises

Classify each triangle in the most precise way possible, given the indicated lengths of its sides and/or measures of its angles.

10.

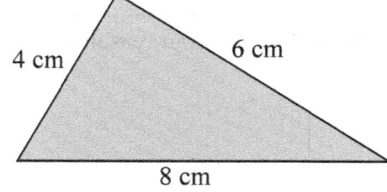

4 cm, 6 cm, 8 cm

11.

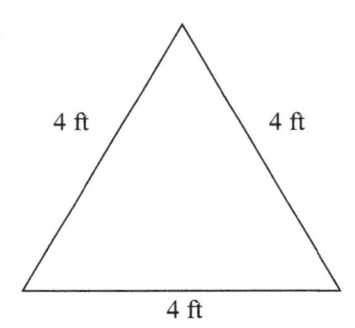

4 ft, 4 ft, 4 ft

Applications

Solve.

12. Suppose the lengths of the sides of △DEF are as shown in the figure. Is this possible? Explain your reasoning.

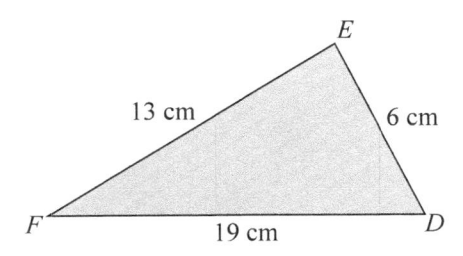

13. In the triangle shown, $m\angle X = 30°$ and $m\angle Y = 70°$.

    a. What is $m\angle Z$?

    b. What kind of triangle is △XYZ?

    c. Which side is opposite $\angle X$?

    d. Which sides include $\angle X$?

    e. Is △XYZ a right triangle?

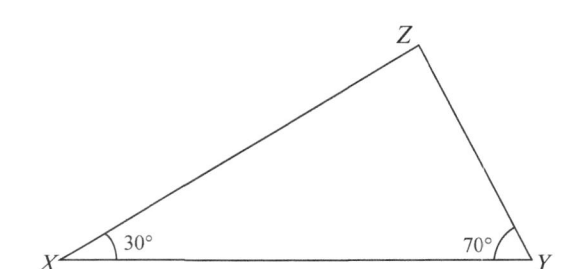

Writing & Thinking

14. Explain, in your own words, the relationships between vertex, ray, angle, and line.

Name: _____ Date: _____ **183**

6.2 Perimeter

Polygon

A **polygon** is a closed plane figure, with _____

Each point where _____

Note: A **closed figure** begins and ends at the same point.

DEFINITION

1. A **triangle** is _____

2. A **parallelogram** is _____
   _____

3. A **rectangle** is _____

4. A **square** is _____

5. A **trapezoid** is _____

Perimeter

The **perimeter** P of a polygon is _____

DEFINITION

Perimeter Formulas for Five Polygons

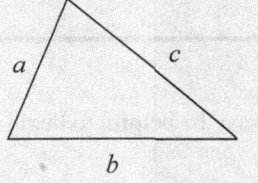

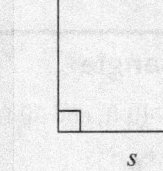

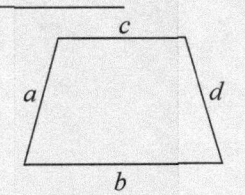

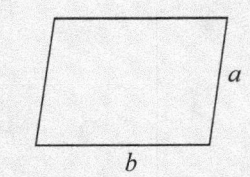

Triangle (sides a, b, c)

$P =$ _____

Square (side s)

$P =$ _____

Rectangle (sides l, w)

$P =$ _____

Trapezoid (sides a, b, c, d)

$P =$ _____

Parallelogram (sides a, b)

$P =$ _____

Note that each formula represents the sum of the lengths of the sides.

FORMULA

184 6.2 Perimeter

▶ Watch and Work

Watch the video for Example 2 in the software and follow along in the space provided.

Example 2 Calculating the Perimeter of a Triangle

Calculate the perimeter of a triangle with sides of length 40 mm, 70 mm, and 80 mm.

Solution

✏ Now You Try It!

Use the space provided to work out the solution to the next example.

Example A Calculating the Perimeter of a Triangle

Calculate the perimeter of a triangle with sides of length 15 ft, 40 ft, and 30 ft. It may be helpful to begin by drawing the figure.

6.2 Exercises

Concept Check

True/False. Determine whether each statement is true or false. If a statement is false, explain how it can be changed so the statement will be true. (**Note:** There may be more than one acceptable change.)

1. a. Every square is a rectangle.

   b. Every rectangle is a square.

2. a. Every parallelogram is a rectangle.

   b. Every rectangle is a parallelogram.

3. A trapezoid has only one pair of parallel lines.

4. The formula to calculate the perimeter of a square is $P = 4s$ where s is the length of one side.

Match each formula for perimeter to its corresponding geometric figure.

5. a. Square   A. $P = 2l + 2w$

   b. Parallelogram   B. $P = 4s$

   c. Rectangle   C. $P = 2b + 2a$

   d. Trapezoid   D. $P = a + b + c$

   e. Triangle   E. $P = a + b + c + d$

Practice

Calculate the perimeter of each figure described.

6. A parallelogram with sides of length 15 cm and 7 cm.

7. A square with sides of length $4\frac{1}{2}$ km.

6.2 Exercises

Calculate the perimeter of each figure.

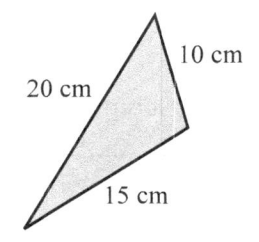

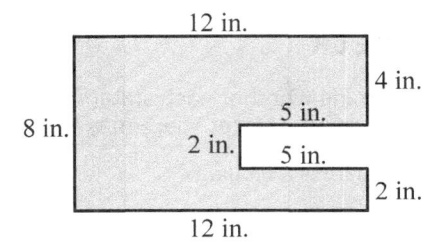

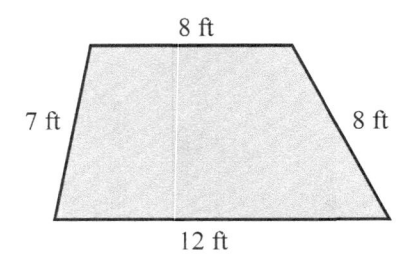

8. 10 cm, 20 cm, 15 cm

10. 12 in., 4 in., 8 in., 5 in., 2 in., 5 in., 2 in., 12 in.

9. 8 ft, 7 ft, 8 ft, 12 ft

Applications

Solve.

11. A five-pointed star-shaped flower plot, where each edge of the star is 6.5 feet, is placed in the middle of a lawn.

  a. What is the perimeter of the plot?

  b. If edging material costs $2.40 per foot, how much will it cost to fully enclose the star?

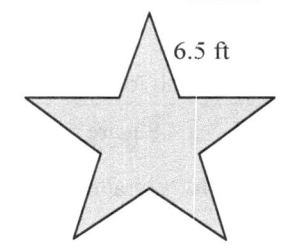

6.5 ft

12. The Pentagon near Washington, D.C., is a five-sided building where each outside wall is 921 feet.

  a. What is the perimeter of the building?

  b. If it takes a person 0.00341 minutes to walk 1 foot, how long will it take the person to walk completely around the building? Round your answer to the nearest tenth of a minute.

Writing & Thinking

13. Explain, briefly, the meaning of perimeter. Write the formula for the perimeter of each of the five types of polygons discussed in this section.

6.3 Area

1. **Area** is a measure of _____.

2. **Area** is measured in _____.

Area Formulas for Five Polygons

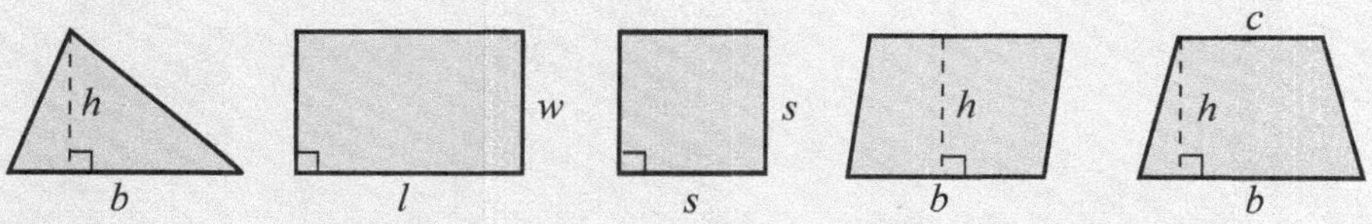

Triangle	Rectangle	Square	Parallelogram	Trapezoid
$A = $ ___	$A = $ ___	$A = $ ___	$A = $ ___	$A = $ ___

Note: The letter h is used to represent the **height** of the figure. The height is also called the **altitude** and is perpendicular to the base.

FORMULA

▶ Watch and Work

Watch the video for Example 2 in the software and follow along in the space provided.

Example 2 Calculating the Area of a Trapezoid Using a Formula

Calculate the area of a trapezoid with altitude 6 in. and parallel sides of length 12 in. and 24 in.

Solution

Now You Try It!

Use the space provided to work out the solution to the next example.

Example A Calculating the Area of a Trapezoid Using a Formula

Calculate the area of a trapezoid with altitude 3 cm and parallel sides of length 9 cm and 15 cm. It may be helpful to begin by drawing the figure.

6.3 Exercises

Concept Check

True/False. Determine whether each statement is true of false. If a statement is false, explain how it can be changed so the statement will be true. (**Note:** There may be more than one acceptable change.)

1. The $(b+c)$ in the trapezoid area formula represents the sum of the lengths of the base and the corners.

2. The height of a triangle is the distance between the base and the vertex opposite the base.

3. The area formula for a triangle is $A = a + b + c$.

4. The area formula for a trapezoid is $A = \frac{1}{2}h(b+c)$.

Practice

Calculate the area of each figure described.

5. A square with sides of length 9 ft.

6. A parallelogram with height 2.3 ft and base 11.9 ft.

Calculate the area of each figure.

7. 8 in., 12 in.

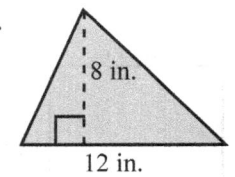

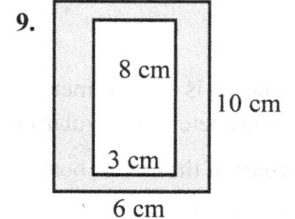

8. 15 yd, 12 yd, 12 yd

9. 8 cm, 3 cm, 10 cm, 6 cm

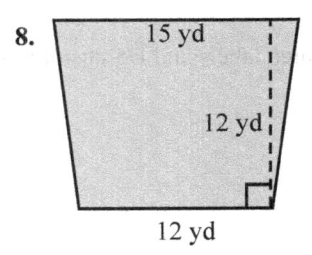

6.3 Exercises

Applications

Solve.

10. The main stage at a theater is in the shape of a trapezoid. The owner of the theater is planning to install a new specially designed flooring system on the stage. The stage is 12 feet wide in the front and 15 feet wide in the back. The stage is 10 feet deep. How much flooring will the manager need?

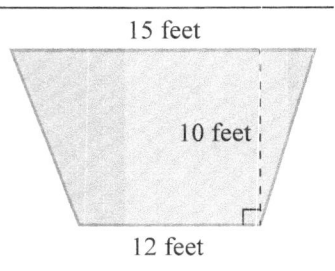

11. A square electronics circuit board is 18 centimeters on each side. On the center of one of the edges is an 8 by 1.5 centimeter rectangular lip for plugging in.
    a. What is the total perimeter of the circuit board, including the lip?
    b. What is the area of the circuit board?

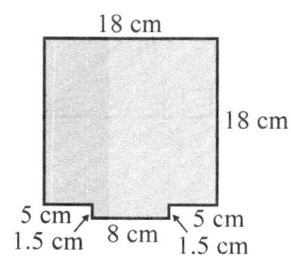

Writing & Thinking

12. Explain why square units are used for labeling areas. Give two examples each of metric area labels and US customary area labels.

13. Explain what the value of $(b+c)$ represents in the formula for the area of a trapezoid.

Name: _____   Date: _____   **191**

6.4 Circles

> **Circles**
>
> **Circle:** The set of all points in a plane that are _____
>
> **Radius:** The distance from the _____
>
> (_____ is used to represent the radius of a circle.)
>
> **Diameter:** The distance from _____
>
> (_____ is used to represent the diameter of a circle and _____.)
>
> **Circumference:** _____
>
> DEFINITION

> **Formulas for Circles**
>
> **Circumference:** $C = $ _____   and $C = $ _____
> **Area:** $A = $ _____
>
> FORMULA

▶ Watch and Work

Watch the video for Example 1 in the software and follow along in the space provided.

Example 1 Calculating the Circumference and Area of a Circle

Calculate **a.** the circumference and **b.** the area of a circle with a radius of 6 ft.

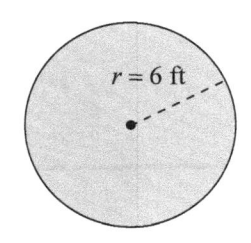

$r = 6$ ft

Solution

✏️ Now You Try It!

Use the space provided to work out the solution to the next example.

Example A Calculating the Circumference and Area of a Circle

Calculate **a.** the circumference and **b.** the area of a circle with a radius of 11 m.

6.4 Exercises

Concept Check

True/False. Determine whether each statement is true of false. If a statement is false, explain how it can be changed so the statement will be true. (**Note:** There may be more than one acceptable change.)

1. Every radius on a circle has the same length.

2. Exact answers can be achieved when substituting 3.14 in place of π.

3. The length of the diameter of a circle is half of the length of the radius.

4. The area of a circle is found by using the formula $A = \pi d$.

Practice

Calculate **a.** the perimeter and **b.** the area of each figure. Use $\pi \approx 3.14$

5.

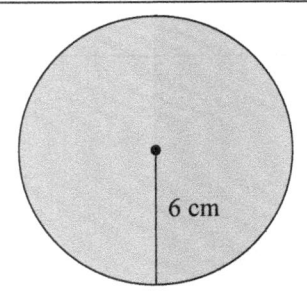

6 cm

6.

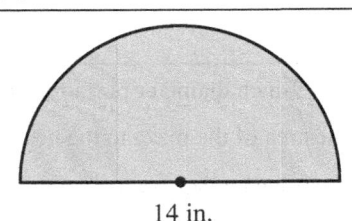

14 in.

194   6.4   Exercises

7.

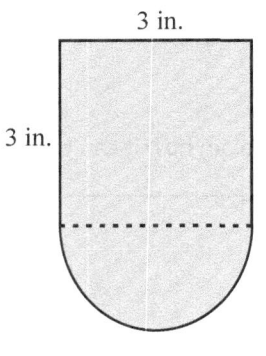

Calculate the area of the shaded portion of the figure. Use $\pi \approx 3.14$.

8.

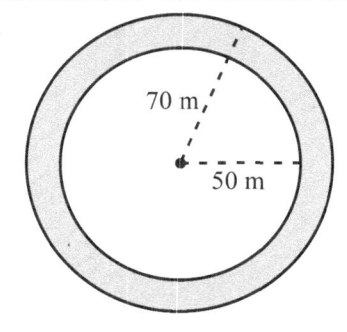

Applications

Solve. Use $\pi \approx 3.14$.

9. Papa Luigi's sells a 9-inch diameter pizza for $8.

   a. Determine the area of the pizza to the nearest tenth.

   b. Determine the price per square inch to the nearest cent per square inch.

10. Rebekah follows a nature trail once around a circular pond. If the distance across the pond (through the center) is 1500 ft, how far did Rebekah walk?

Writing & Thinking

11. Explain why $2\pi r$ is equivalent to πd.

12. Propose a method for calculating the area of a semicircle and justify your method.

Name: _____ Date: _____ **195**

6.5 Volume and Surface Area

1. **Volume** is a measure of the _____

2. Volume is measured in _____

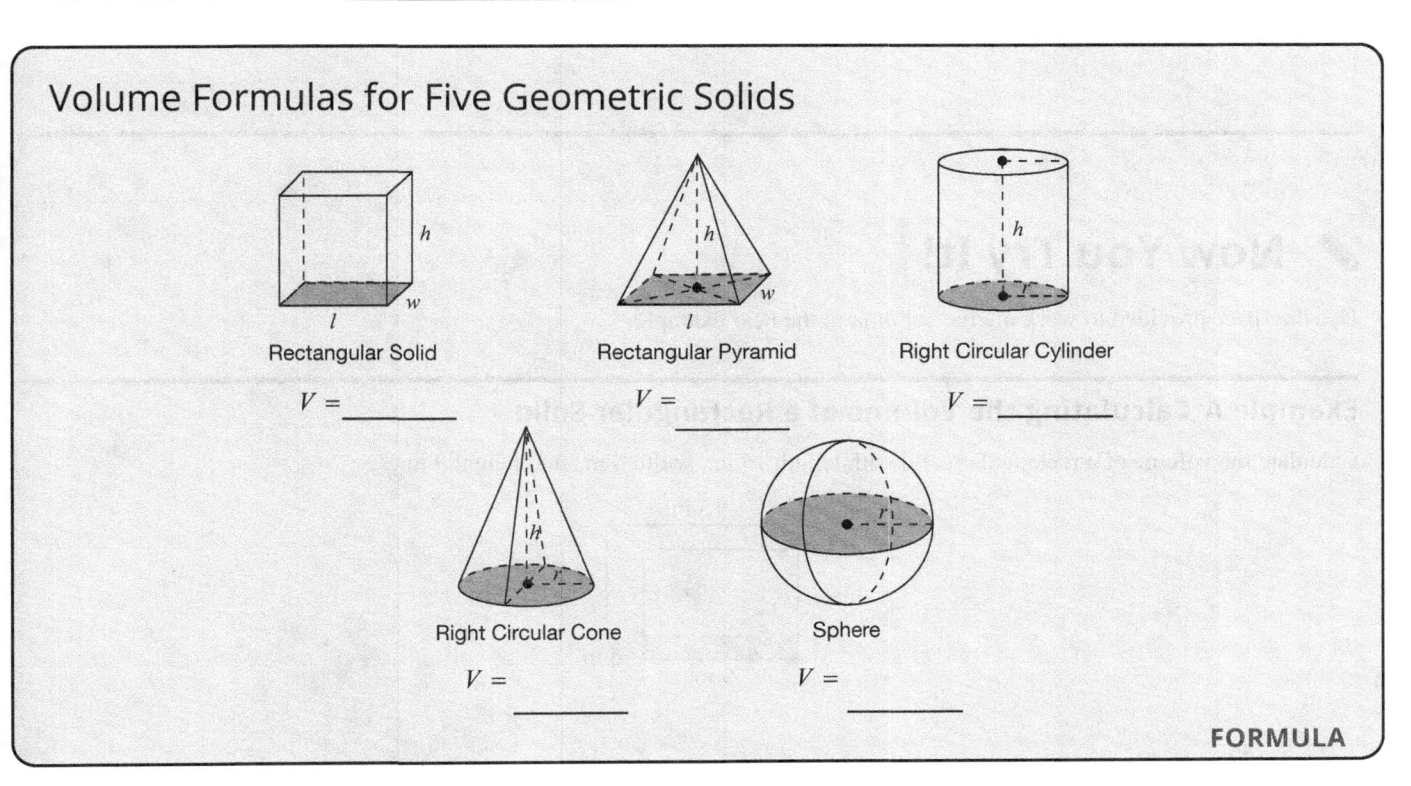

Volume Formulas for Five Geometric Solids

Rectangular Solid
$V = $ _____

Rectangular Pyramid
$V = $ _____

Right Circular Cylinder
$V = $ _____

Right Circular Cone
$V = $ _____

Sphere
$V = $ _____

FORMULA

▶ Watch and Work

Watch the video for Example 1 in the software and follow along in the space provided.

Example 1 Calculating the Volume of a Rectangular Solid

Calculate the volume of a rectangular solid with length 8 in., width 4 in., and height 12 in.

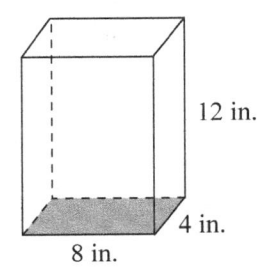

12 in.
4 in.
8 in.

6.5 Volume and Surface Area

Solution

✏️ Now You Try It!

Use the space provided to work out the solution to the next example.

Example A Calculating the Volume of a Rectangular Solid

Calculate the volume of a rectangular solid with length 15 in., width 6 in., and height 9 in.

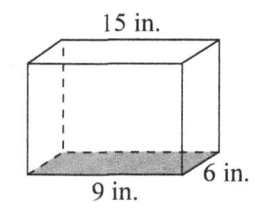

---

3. The **surface area** (*SA*) of a geometric solid is _____

Surface Area Formulas for Three Geometric Solids

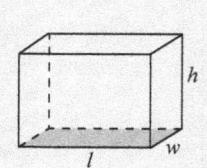

Rectangular Solid
SA = _____

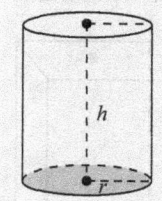

Right Circular Cylinder
SA = _____

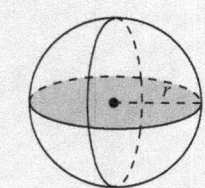

Sphere
SA = _____

FORMULA

6.5 Exercises

Concept Check

True/False. Determine whether each statement is true or false. If a statement is false, explain how it can be changed so the statement will be true. (**Note:** There may be more than one acceptable change.)

1. To find the volume of a can of corn, the formula $V = \pi r^2 h$ would be used.

2. $V = lwh$ is the formula for the surface area of a rectangular solid.

3. The area of the paper label on a can of peaches is an example of surface area.

4. To find the volume of a rectangular solid, the areas of each surface are added together.

Match each formula for volume to its corresponding geometric figure.

5. a. Rectangular solid   A. $V = \dfrac{4}{3}\pi r^3$

   b. Rectangular pyramid   B. $V = \dfrac{1}{3}\pi r^2 h$

   c. Right circular cylinder   C. $V = lwh$

   d. Right circular cone   D. $V = \pi r^2 h$

   e. Sphere   E. $V = \dfrac{1}{3}lwh$

Practice

Calculate the volume of each solid. Use $\pi \approx 3.14$.

6. A rectangular solid with length 5 in., width 2 in., and height 7 in.

7. A right circular cone 3 mm high with a 2 mm radius.

6.5 Exercises

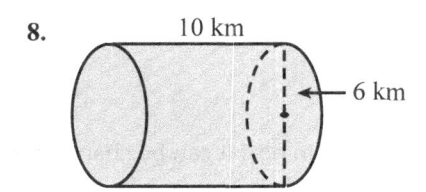

8. 10 km, 6 km

Calculate the surface area of each solid. Use π ≈ 3.14.

9. 9 mm

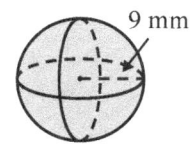

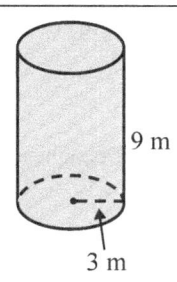

10. 9 m, 3 m

Applications

Solve. Use π ≈ 3.14.

11. A rectangular tent with straight sides has a pyramidal shaped roof. The dimensions of the rectangular portion are 12 ft long, 10 ft wide, and 6 ft high. The peak of the pyramid is 2 ft above the top edge of the walls. What is the volume of the inside of the tent?

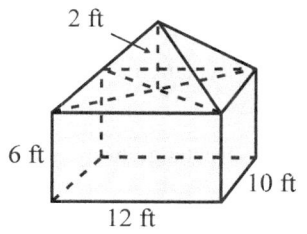

12. Disposable paper drinking cups like those used at water coolers are often cone-shaped. Find the volume of such a cup that is 9 cm high with a 3.2 cm radius. Express the answer to the nearest milliliter.

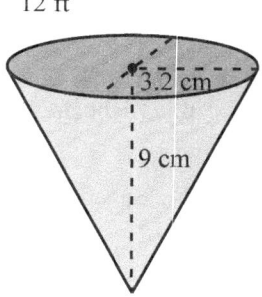

Writing & Thinking

13. Discuss the type of units used for volume and explain why.

14. List the steps and formulas you would use to find the volume of an ice cream cone (assuming the ice cream itself forms a perfect half sphere).

6.6 Similar and Congruent Triangles

Similar Triangles

1. In similar triangles, the corresponding _____

2. In similar triangles, the _____

DEFINITION

Properties of Congruent Triangles

Two triangles are congruent if:

1. _____

2. _____

PROPERTIES

Determining Congruent Triangles

1. **Side-Side-Side (SSS)**

   If two triangles are such that the _____
   _____

2. **Side-Angle-Side (SAS)**

   If two triangles are such that the _____
   _____

3. **Angle-Side-Angle (ASA)**

   If two triangles are such that _____
   _____
   _____

DEFINITION

▶ Watch and Work

Watch the video for Example 4 in the software and follow along in the space provided.

Example 4 Determining Whether Triangles are Congruent

Determine whether triangles *PQR* and *MNO* are congruent.

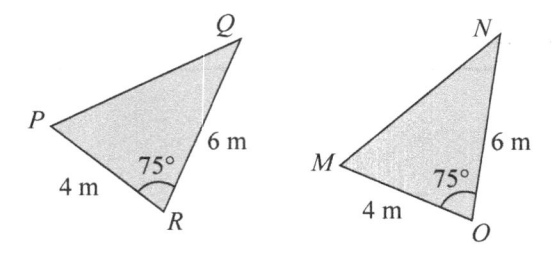

Solution

✏ Now You Try It!

Use the space provided to work out the solution to the next example.

Example A Determining Whether Triangles are Congruent

Determine whether triangles *JKL* and *MNO* are congruent. If they are congruent, state the property that confirms that they are congruent.

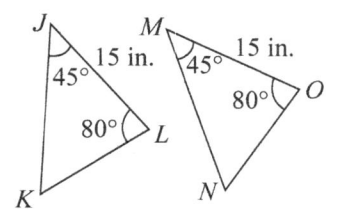

6.6 Exercises

Concept Check

True/False. Determine whether each statement is true or false. If a statement is false, explain how it can be changed so the statement will be true. (**Note:** There may be more than one acceptable change.)

1. Similar triangles have corresponding sides that are equal.

2. If $\triangle ABC \cong \triangle DEF$, then the measure of angle C equals the measure of angle D.

3. If $\triangle ABC \sim \triangle DEF$, then $AC = DF$.

4. Congruent triangles have corresponding angles that are equal.

Practice

Determine whether each pair of triangles is similar. If the pair of triangles is similar, explain why and indicate the similarity by using the ~ symbol.

5.

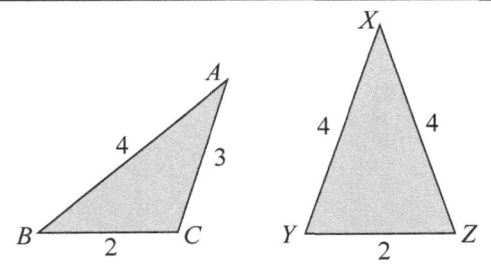

6.

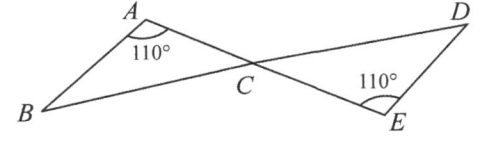

6.6 Exercises

Find the values for *x* and *y*.

7. △ABC ~ △XYZ

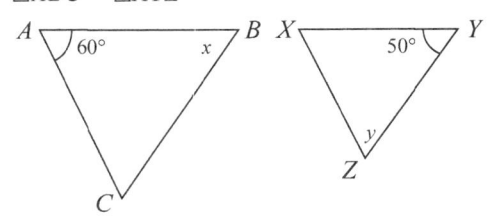

8. △ABE ~ △CDE

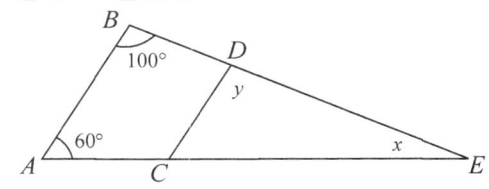

Determine whether each pair of triangles is congruent. If the pair of triangles is congruent, state the property that confirms that they are congruent.

9.

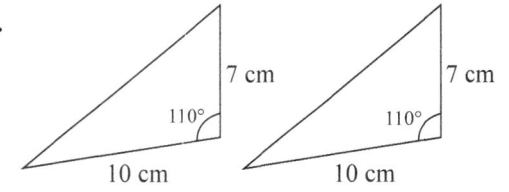

10.

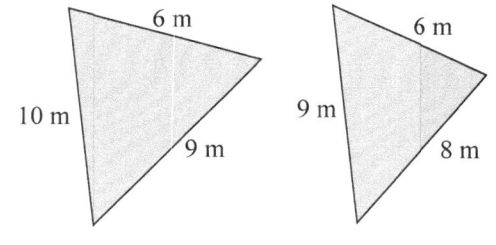

Applications

Solve.

11. A child's playhouse is built to look like a smaller version of the family house, where the ends of the roofs have similar proportions. The width of the main house (*AB*) is 32 feet and the length from the peak to the gutter of the roof for one of the sides is 20 feet. If the width of the playhouse (*DF*) is 12 feet, what is the length from the peak to the gutter (*DE*) of the playhouse roof?

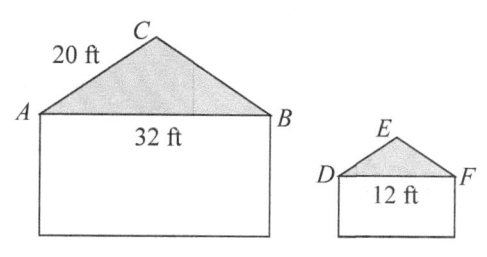

12. Your neighbors are hanging their holiday lights. The ladder they are currently using is 12 feet long and when leaned up against the house just reaches the top of their 8-foot tall porch. How long of a ladder will they need to reach the top of their chimney which is at a height of 32 feet? (Assume that both ladders are placed such that they make the same angle with the ground.)

Writing & Thinking

13. Determine the errors in the following statement. Assume $\triangle ABC \sim \triangle DEF$.

    a. Corresponding angles are congruent. This means $m\angle A = m\angle D$, $m\angle B = m\angle F$, and $m\angle C = m\angle E$.

    b. Corresponding sides are the same length.

Name: Date:

6.7 Square Roots and the Pythagorean Theorem

Terminology of Radicals

The symbol $\sqrt{}$ is called _____

The number under the radical sign is called _____

The complete expression, such as $\sqrt{49}$, is called _____

DEFINITION

▶ Watch and Work

Watch the video for Example 2 in the software and follow along in the space provided.

Example 2 Evaluating Square Roots

Use your memory of the values in Table 2 to evaluate each expression.

a. $\sqrt{256}$

b. $\sqrt{81}$

Solution

✏ Now You Try It!

Use the space provided to work out the solution to the next example.

Example A Evaluating Square Roots

Evaluate each expression.

a. $\sqrt{36}$

b. $\sqrt{169}$

Terms Related to Right Triangles

Right triangle: _____

Hypotenuse: _____

Leg: _____

DEFINITION

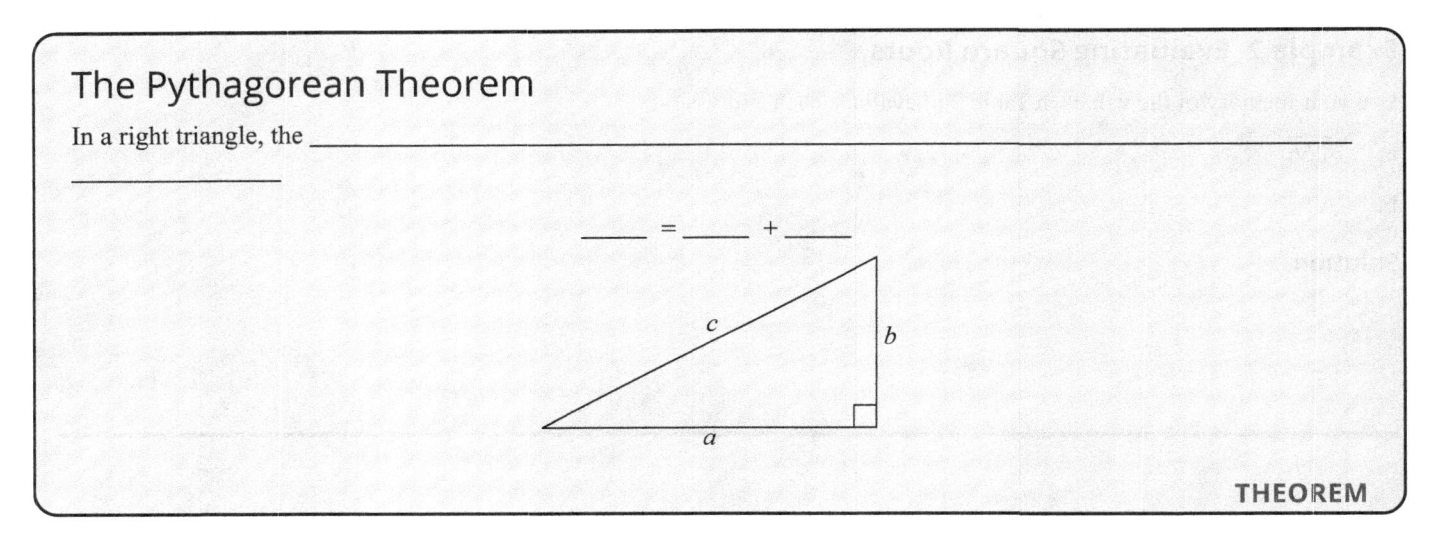

The Pythagorean Theorem

In a right triangle, the _____
_____

____ = ____ + ____

THEOREM

6.7 Exercises

Concept Check

True/False. Determine whether each statement is true or false. If a statement is false, explain how it can be changed so the statement will be true. (**Note:** There may be more than one acceptable change.)

1. 49 is a perfect square.

2. In the expression $\sqrt{81}$, the number 9 is the radicand.

3. The Pythagorean Theorem can be used to find the length of the longest side of a right triangle if the lengths of the two legs are known.

4. The Pythagorean Theorem works for any type of triangle.

Practice

Evaluate each expression.

5. $\sqrt{36}$

6. $\sqrt{225}$

Use the Pythagorean Theorem to determine whether or not each triangle is a right triangle.

7. 8 in., 6 in., 10 in.

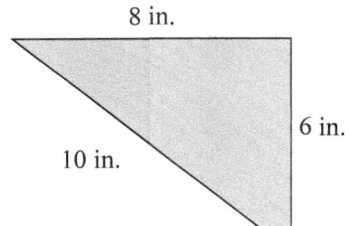

8. 6 yd, 14 yd, 11 yd

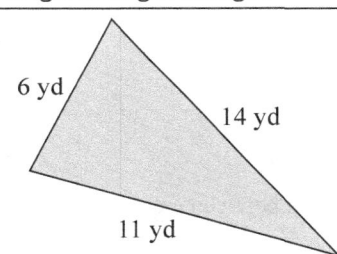

Find the hypotenuse for each right triangle accurate to the nearest hundredth.

9. 4, 3, c

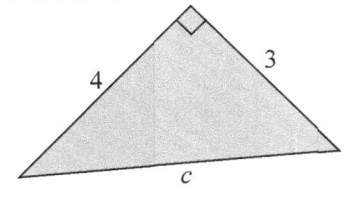

Applications

Solve.

10. The base of a fire engine ladder is 30 feet from a building and reaches to a third floor window 50 feet above ground level. Find the length of the ladder to the nearest hundredth of a foot.

11. The shape of home plate in the game of baseball can be created by cutting off two triangular pieces at the corners of a square, as shown in the figure. If each of the triangular pieces has a hypotenuse of 12 inches and legs of equal length, what is the length of one side of the original square, to the nearest tenth of an inch?

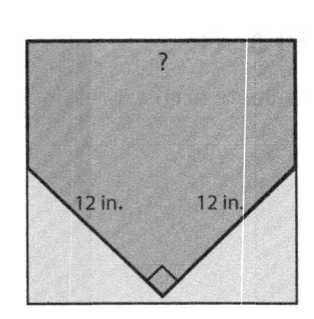

Writing & Thinking

12. Explain the connection between a perfect square and its square root. Give an example.

Chapter 6 Project

Before and After

An activity to demonstrate the use of geometric concepts in real life.

Suppose HGTV came to your home one day and said, "Congratulations, you have just won a FREE makeover for any room in your home! The only catch is that you have to determine the amount of materials needed to do the renovations and keep the budget under $2000." Could you pass up a deal like that? Would you be able to calculate the amount of flooring and paint needed to remodel the room? Remember it's a FREE makeover if you can!

Let's take an average size room that is rectangular in shape and measures 16 feet 3 inches in width by 18 feet 9 inches in length. The height of the ceiling is 8 feet. The plan is to repaint all the walls and the ceiling and to replace the carpet on the floor with hardwood flooring. You are also going to put crown molding around the top of the walls for a more sophisticated look.

1. Take the length and width measurements that are in feet and inches and convert them to a fractional number of feet and reduce to lowest terms. (Remember that there are 12 inches in a foot. For example, 12 feet 1 inch is $12\frac{1}{12}$ feet.)

2. Now convert these same measurements to decimal numbers.

3. Determine the number of square feet of flooring needed to redo the floor. (Express your answer in terms of a decimal and do not round the number.)

4. If the flooring comes in boxes that contain 24 square feet, how many boxes of flooring will be needed? (Remember that the store only sells whole boxes of flooring.)

5. If the flooring you chose costs $74.50 per box, how much will the hardwood flooring for the room cost (before sales tax)?

6. Figure out the surface area of the four walls and the ceiling that need to be painted, based on the room's dimensions. (We will ignore any windows, doors, or closets since this is an estimate.)

7. Assume that a gallon of paint covers 350 square feet and you are going to have to paint the walls and the ceilings **twice** to cover the current paint color. Determine how many gallons of paint you need to paint the room. (Again, assume that you can only buy whole gallons of paint. Any leftover paint can be used for touch-ups.)

8. If the paint you have chosen costs $18.95 per gallon, calculate the cost of the paint (before sales tax).

9. Determine how many feet of crown molding will be needed to go around the top of the room.

10. The molding comes in 12-foot sections only. How many sections will you need to buy?

11. If the molding costs $2.49 per linear foot, determine the cost of the molding (before sales tax).

12. Calculate the cost of all the materials for the room makeover (before sales tax).

    a. Were you able to stay within budget for the project?

    b. If so, then what extras could you add? If not, what could you adjust in this renovation to stay within budget?

    c. Using sales tax in your area, calculate the final price of the room makeover with sales tax included.

Chapter 6 Project

This Mixtape Is Fire!

An activity to demonstrate the use of geometric formulas and properties in real life.

While most music listening today happens via online streaming, music is still distributed in a variety of physical formats, such as cassette tapes, CDs, and vinyl records.

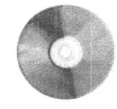

Cassette Tape   Compact Disc (CD)   Vinyl Record

Imagine that you are a musical artist who wants to produce and distribute your own music using cassette tapes, CDs, and vinyl records. To save money, you plan to design the album artwork yourself and send the art files to the manufacturer along with the music files. The album artwork contains the cover art, information about each song, and all the credits. The dimensions of the album artwork for each type of physical media varies.

1. For the cassette tape, the album artwork (called a J-card) is commonly printed on a 4-inch by 4.11-inch rectangle that is folded into a J shape and placed into a plastic case.
   a. What is the perimeter of this sleeve?
   b. What is the area of this sleeve?

2. The plastic case for a cassette tape commonly measures 4.25 inches by 2.75 inches by 0.6 inches. The case is in the shape of a rectangular prism.
   a. What is the surface area of a cassette case?
   b. What is the volume of a cassette case? Round to the nearest hundredth.

3. For the CD, the front artwork is commonly printed onto a 4.724-inch square and the back artwork is printed onto a 4.724-inch by 5.96-inch rectangle. Both the front and back artwork is inserted into a plastic case with dimensions 4.91 inches by 5.61 inches by 0.39 inches. The case is in the shape of a rectangular prism. The entire case is then wrapped in plastic film.
   a. What is the total area of the front and back artwork? Round to the nearest hundredth.
   b. What is the volume of the CD case? Round to the nearest hundredth.
   c. A less common CD case has dimensions 4.92 inches by 5.51 inches by 0.41 inches. Which CD case size would require less plastic film to wrap it? How do you know?

4. The vinyl record sleeve has a 12.375-inch square for the cover art. Suppose you want to designate a triangle at the top right of the cover art for the artist's name. This triangle has one leg along the right side of the sleeve measuring 3 inches and another leg along the top side of the sleeve measuring 4 inches.
   a. What is the perimeter of the cover art excluding the triangle? Explain how you found this perimeter, using the Pythagorean Theorem in your explanation.
   b. What is the area of the cover art excluding the triangle? Round to the nearest hundredth.

5. Each vinyl record is circular with a diameter of 12 inches. Use the approximation $\pi \approx 3.14$.
   a. What is the circumference of a 12-inch vinyl record?
   b. In the middle of each record, there is a circle with a diameter of 3 inches with no music on it where a paper label is placed to identify the record. What is the area of the playable surface of this vinyl record that is not the center label? Explain how you found your answer.
   c. You plan to have 500 vinyl records made for your album. At their center, each record has a height of 0.3 inches. The height along the outside of the record is 0.2 inches with a rounded edge. Can you use the formula for a volume of a cylinder to calculate the exact volume of a stack of 500 records? Explain your answer.

CHAPTER 7

Statistics, Graphs, and Probability

7.1 Statistics: Mean, Median, Mode, and Range

7.2 Reading Graphs

7.3 Constructing Graphs from Databases

7.4 Probability

CHAPTER 7 PROJECTS
What's My Average?

Misleading Graphs

Connections

From data in reports at work to news from media outlets, people come across statistics and probability daily. A lot of the statistics calculated about the US population are the result of the data collected in the US Census, which is updated every 10 years. Besides being used in news and research, these statistics can be used by businesses to determine the best way to make money long term.

The field of actuary science uses statistics to help insurance companies determine how much to charge customers. For example, your insurance premium is based on the likelihood of you filing a claim based on personal information, such as your age, your gender, and where you live. These companies base their rates on statistics gathered over many years. Your age affects insurance rates because drivers under the age of 25 on average have a higher rate of fatal accidents. Similarly, your location can affect your insurance rate since accident rates vary by location. For example, Fort Collins, CO, has a lower accident rate per driver than Portland, OR. Knowing how to interpret statistics and probability will help you make and understand decisions that affect your everyday life.

Suppose you are employed at the local city planning department. Over the past three-week period, or 15 business days, the department issued the following number of building permits to residents and contractors.

17, 19, 18, 35, 30, 29, 23, 14, 18, 16, 20, 18, 18, 25, 30

For the monthly review meeting, you need to present this data in a meaningful way. How would you calculate the mean, median, mode, and range of this data? What do these measurements tell you about the number of building permits issued?

7.1 Statistics: Mean, Median, Mode, and Range

1. **Statistics** is the study of how to _____.

2. A **statistic** is a particular measure or characteristic of a part, or _____, of a larger collection of items called the _____.

Terms Used in the Study of Statistics

_____	Value(s) measuring some characteristic of interest such as income, height, weight, grade point averages, scores on tests, and so on. (We will consider only numerical data.)
_____	A single number describing some characteristic of the data.
_____	The sum of all the data divided by the number of data items. (The arithmetic average of the data.)
_____	The middle of the data after the data have been arranged in order (smallest to largest or vice versa). (The median may or may not be one of the data items.)
_____	The data item(s) that appears the most number of times. (A set of data may have more than one mode.)
_____	The difference between the largest and smallest data items.

DEFINITION

▶ Watch and Work

Watch the video for Example 1 in the software and follow along in the space provided.

Example 1 Application: Finding the Mean

Find the mean annual income for the individuals in Group A.

Group A
Annual Income for 8 Individuals

$28,000	$45,000	$22,000	$80,000	$25,000	$25,000	$27,000	$30,000

Table 1

7.1 Statistics: Mean, Median, Mode, and Range

Solution

✏️ Now You Try It!

Use the space provided to work out the solution to the next example.

Example A Application: Finding the Mean

Find the mean movie time for the movies in Group B, rounded to the nearest minute.

Group B					
The Time (in Minutes) of 11 Movies					
100 min	90 min	113 min	110 min	88 min	90 min
155 min	88 min	105 min	93 min	90 min	

Table 2

To Find the Median

1. Arrange the data in _____
2. a. If there is an odd number of items, _____
   b. If there is an even number of items, _____
      (**Note:** This value _____.)

PROCEDURE

7.1 Exercises

Concept Check

True/False. Determine whether each statement is true or false. If a statement is false, explain how it can be changed so the statement will be true. (**Note:** There may be more than one acceptable change.)

1. Data is the value(s) of a particular characteristic of interest such as number of people, weight, temperatures, or innings pitched.

2. The range is the difference between the first number listed in the data and the last number listed.

3. The number that appears the greatest number of times in a set of data is the sample.

4. The mean and median are never the same number.

Practice

For each set of data, find **a.** the mean, **b.** the median, **c.** the mode (if any), and **d.** the range.

5. The ages of the first five US presidents on the date of their inaugurations were as follows. (The presidents were Washington, Adams, Jefferson, Madison, and Monroe.)

   57, 61, 57, 57, 58

6. Dr. Wright recorded the following nine test scores for students in his statistics course.

   95, 82, 85, 71, 65, 85, 62, 77, 98

7. Family incomes in a survey of eight students are as follows.

   $35,000, $63,000, $28,000, $36,000, $42,000, $51,000, $71,000, $63,000

8. Stacey went to six different auto repair shops to get the following estimates to repair her car.

   $425, $525, $325, $300, $500, $325

7.1 Exercises

Applications

Solve.

9. Suppose that you are to take four hourly exams and a final exam in your chemistry class. Each exam has a maximum of 100 points and you must average between 75 and 82 points to receive a passing grade of C. If you have scores of 83, 65, 70, and 78 on the hourly exams, what is the minimum score you can make on the final exam and receive a grade of C? (First, explain your strategy in solving this problem. Then, solve the problem.)

10. The number of farms in the United States is decreasing. The following graph shows the number of farms (in millions) for each decade since 1940. As you can tell from the graph, the number of farms seems to be leveling off somewhat.

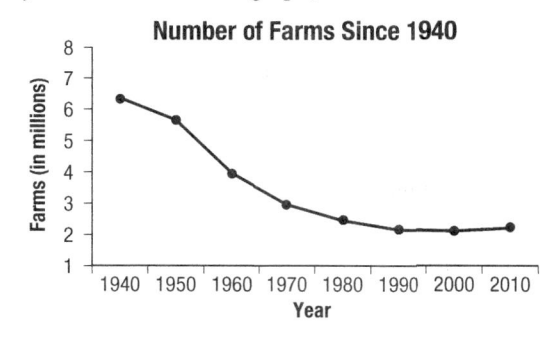

    Number of Farms Since 1940

1940: 6.35	1950: 5.65
1960: 3.96	1970: 2.95
1980: 2.44	1990: 2.15
2000: 2.11	2010: 2.2

    a. Find the mean number of farms from 1940 to 2010.

    b. Find the mean number of farms from 1980 to 2010.

Writing & Thinking

11. Determine whether or not the mean and median represent the same number. Give examples to justify your answer.

12. Give three different and specific examples where some statistical measure is used (outside of a class).

Name: _____ Date: _____ **219**

7.2 Reading Graphs

Four Types of Graphs and Their Purposes

_____	To emphasize comparative amounts
_____	To help in understanding percents or parts of a whole (_____ are also called pie charts.)
_____	To indicate tendencies or trends over a period of time
_____	To indicate data in classes (a range or interval of numbers)

DEFINITION

Properties of Graphs

Every graph should

1. _____
2. _____
3. _____

PROPERTIES

▶ Watch and Work

Watch the video for Example 2 in the software and follow along in the space provided.

Example 2 Reading a Circle Graph

Examine the circle graph. This graph shows the percent of a household's annual income they plan to budget for various expenses. Suppose the household has an annual income of $45,000. Use the information in the graph to calculate how much money will be budgeted for each expense.

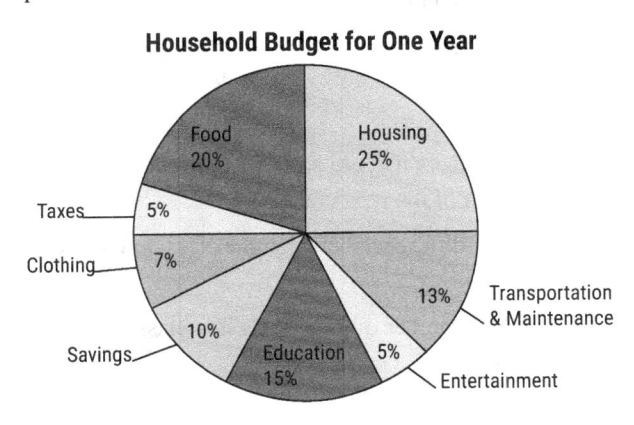

Household Budget for One Year

- Food 20%
- Housing 25%
- Taxes 5%
- Clothing 7%
- Savings 10%
- Education 15%
- Entertainment 5%
- Transportation & Maintenance 13%

7.2 Reading Graphs

Solution

✏️ Now You Try It!

Use the space provided to work out the solution to the next example.

Example A Reading a Circle Graph

Using the circle graph in Example 2, how much will the family spend on each of the following expenses if the family income increases to $55,000 and the percent budgeted for each expense does not change?

    a.   Housing

    b.   Savings

    c.   Clothing

    d.   Food

Terms Related to Histograms

_____	A range (or interval) of numbers that contains data items.
_____	The smallest number that belongs to a class.
_____	The largest number that belongs to a class.
_____	Numbers that are halfway between the upper limit of one class and the lower limit of the next class.
_____	The difference between the class boundaries of a class (the width of each bar).
_____	The number of data items in a class.

DEFINITION

7.2 Exercises

Concept Check

True/False. Determine whether each statement is true or false. If a statement is false, explain how it can be changed so the statement will be true. (**Note:** There may be more than one acceptable change.)

1. Graphs should always be clearly labeled, easy to read, and have appropriate titles.

2. Circle graphs show trends over a period of time.

3. The frequency is the number of data items in a class.

4. Numbers that are halfway between the upper limit of one class and the lower limit of the next class are the class boundaries.

Applications

Answer the questions using the given graphs.

5. The following bar graph shows the number of students in five fields of study at a university.

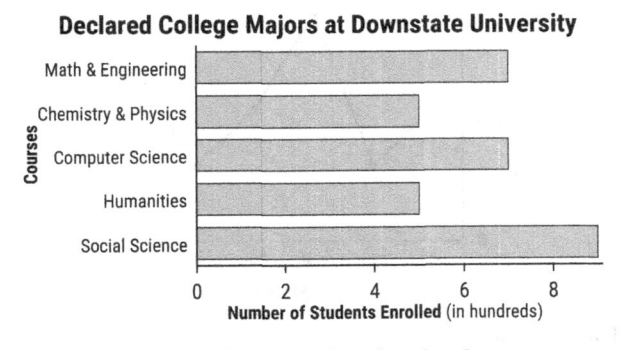

Declared College Majors at Downstate University

Courses (from top to bottom): Math & Engineering (~7), Chemistry & Physics (~5), Computer Science (~7), Humanities (~5), Social Science (~9)

Number of Students Enrolled (in hundreds)

a. Which field(s) of study has the largest number of declared majors?

b. Which field(s) of study has the smallest number of declared majors?

c. How many declared majors are indicated in the entire graph?

d. What percent are computer science majors? Round your answer to the nearest tenth of a percent.

6. The following circle graph represents the various areas of spending for a school with a total budget of $34,500,000.

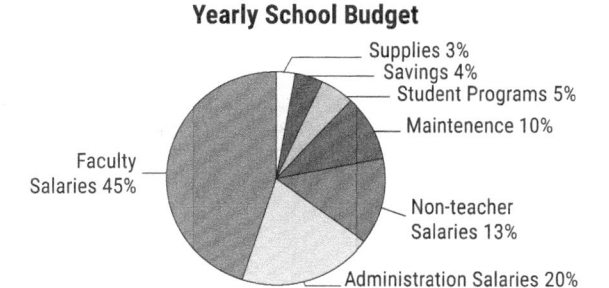

Yearly School Budget

- Supplies 3%
- Savings 4%
- Student Programs 5%
- Maintenence 10%
- Non-teacher Salaries 13%
- Administration Salaries 20%
- Faculty Salaries 45%

a. What amount will be allocated to each category?

b. What percent will be for expenditures other than salaries?

c. How much will be spent on maintenance and supplies

d. How much more will be spent on teachers' salaries than on administration salaries?

7. The following line graph shows the total monthly rainfall in a certain city over five months.

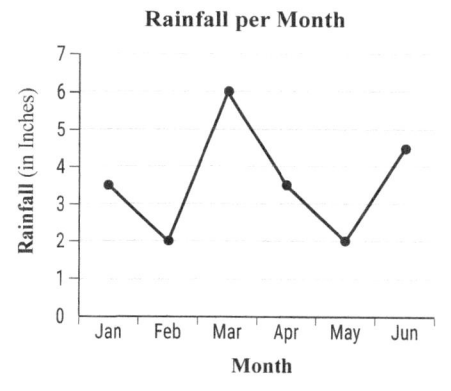

Rainfall per Month

a. Which months had the least rainfall?

b. What was the most rainfall in a month?

c. What month had the most rainfall?

d. What was the mean rainfall over the six-month period (to the nearest hundredth)?

8. The following histogram summarizes the tread life for 100 types of new tires.

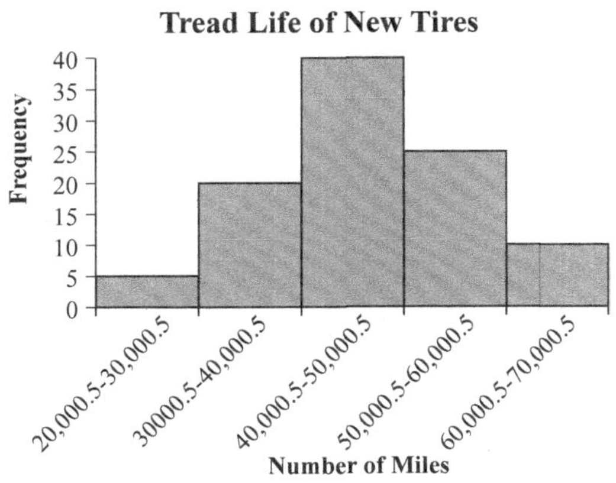
Tread Life of New Tires
Frequency
40
35
30
25
20
15
10
5
0
20,000.5-30,000.5
30000.5-40,000.5
40,000.5-50,000.5
50,000.5-60,000.5
60,000.5-70,000.5
Number of Miles

a. How many classes are represented?

b. What is the width of each class?

c. Which class has the highest frequency?

Writing & Thinking

9. State three properties or characteristics that should be true of all graphs so that they can communicate numerical data quickly and easily.

10. Compare and contrast a bar graph and a histogram.

7.2 Exercises

7.3 Constructing Graphs from Databases

Steps to Follow in Constructing a Vertical Bar Graph
1. Draw a vertical axis and _____
2. Mark an appropriate scale on the _____
   (The scale must _____.)
3. Mark the categories of _____.
4. Draw the vertical bar for each category so that the _____
   _____.
   (Note: Make sure the bars _____.)
5. Give the graph _____.

PROCEDURE

Steps to Follow in Constructing a Circle Graph
1. Find the central angle (angle at the center of the circle) for each category by _____
   _____
2. _____
3. _____
4. Label each sector with the _____
5. Give the graph _____.

PROCEDURE

7.3 Exercises

Concept Check

True/False. Determine whether each statement is true or false. If a statement is false, explain how it can be changed so the statement will be true. (**Note:** There may be more than one acceptable change.)

1. In creating a vertical bar graph, a bar's width should vary based on the number it represents.

2. The first step in constructing a vertical bar graph is to draw a vertical and a horizontal axis.

3. Bar graphs can have either vertical or horizontal bars.

4. Titles are unnecessary for circle graphs.

7.3 Exercises

Applications

For each set of data, construct the specified graph.

5. Construct a bar graph that represents the following data.

Largest Islands of the World

Island	Area in Square Miles (nearest ten thousand)
Greenland	840,000
New Guinea	310,000
Borneo	290,000
Madagascar	230,000
Baffin	200,000
Sumatra	180,000
Honshu	90,000
Great Britain	90,000

6. Construct a bar graph that represents the following data.

10 Top Films by Domestic Box Office Earnings, 2020

Motion Picture	Box Office Gross (in millions of dollars)
Bad Boys for Life	204.4
1917	157.9
Sonic the Hedgehog	146.1
Jumanji: The Next Level	124.7
Star Wars: Episode IX - The Rise of Skywalker	124.5
Birds of Prey	84.2
Dolittle	77.0
Little Women	70.5
The Invisible Man	64.9
The Call of the Wild	62.3

Source: Information courtesy of Box Office Mojo. Used with permission. www.boxofficemojo.com

7. Construct a circle graph that represents the following data.

Percent of Population with Particular Blood Types

Type of Blood	Percent of Population
O positive (O⁺)	38%
O negative (O⁻)	7%
A positive (A⁺)	34%
A negative (A⁻)	6%
B positive (B⁺)	9%
B negative (B⁻)	2%
AB positive (AB⁺)	3%
AB negative (AB⁻)	1%

Source: AABB.org

8. Construct a circle graph that represents the following data.

Global Share of Total Energy Supply by Source, 2019

Source of Energy	Percent
Oil	30.9%
Coal	26.8%
Natural Gas	23.2%
Biofuels	9.4%
Nuclear	5.0%
Hydro	2.5%
Other	2.2%

Source: Key World Energy Statistics, 2021. International Energy Agency.

Writing & Thinking

9. While most graphs can be created through the use of a computer, give at least one benefit from constructing a graph yourself.

10. List three mistakes a person might make when constructing a graph.

Name: _____ Date: _____ **229**

7.4 Probability

1. Activities involving chance such as tossing a coin, rolling a die, spinning a wheel in a game, and predicting weather are called _____. The likelihood of a particular result is called its _____.

Terms Related to Probability

_____	An individual result of an experiment.
_____	The set of all possible outcomes of an experiment.
_____	Some (or all) of the outcomes from the sample space.

DEFINITION

2. A **tree diagram** can be used to _____. Each branch of the tree diagram shows _____.

▶ Watch and Work

Watch the video for Example 3 in the software and follow along in the space provided.

Example 3 Finding the Sample Space Using a Tree Diagram

A coin is tossed and then one of the numbers (1, 2, and 3) is chosen at random from a box. Draw a tree diagram illustrating the possible outcomes of the experiment and list the outcomes in the sample space.

Solution

230   7.4   Probability

✏️ Now You Try It!

Use the space provided to work out the solution to the next example.

Example A Finding the Sample Space Using a Tree Diagram

A coin is tossed and then a six-sided die is rolled. Draw a tree diagram illustrating the outcomes of the experiment and list the outcomes in the sample space.

Probability of an Event

$$\text{probability of an event} = \underline{}$$

DEFINITION

Basic Characteristics of Probabilities.

1. Probabilities are between 0 and 1, inclusive.
   a. If an event can never occur, _____
   b. If an event will always occur, _____
2. The sum of the probabilities of the outcomes in a sample space is 1.

PROPERTIES

7.4 Exercises

Concept Check

True/False. Determine whether each statement is true or false. If a statement is false, explain how it can be changed so the statement will be true. (**Note:** There may be more than one acceptable change.)

1. The individual result of an experiment is a probability.

2. An event is some or all of the outcomes from the sample space.

3. A single result of an experiment is an outcome.

4. The probability of a tossed coin showing either heads or tails is 1.

Applications

For each experiment, draw a tree diagram illustrating the possible outcomes and list the outcomes in the sample space.

5. Four marbles are in a box: one red, one white, one blue, and one purple. One marble is chosen.

6. There are three flavors of potato chips to choose from: original, BBQ, and cheddar. One flavor is chosen.

7.4 Exercises

For each problem, calculate the probability described.

7. A box contains 5 marbles: two red, one white, two blue. What is the probability of choosing a blue marble from the box?

8. A machine contains only 5 gumballs: three yellow, one white, one green. What is the probability of getting a yellow gumball when you put a coin in the machine?

Writing & Thinking

9. List at least three activities that are experiments of chance.

10. Explain the benefit(s) of using a tree diagram to determine probability.

Chapter 7 Project

What's My Average?
An activity to investigate the relationships among different measures of central tendency.

If you are a college student, then grades are important to you. They determine whether you are eligible for scholarships or getting into a particular college or program of choice. It is important to be able to calculate your grade point average in a class and to be able to determine the score you need on a test to reach your desired average. Professors have many different ways of calculating your average for a class. Measures of average are often referred to as measures of central tendency.

For this project, you will be working with two of these measures, the **mean** and the **median**.

Recall that the **mean** of a set of data is found by adding all the numbers in the set and then dividing by the number of data values. The **median** is the middle number once you arrange the data in order from smallest to largest. If there is an even number of data values, then the median is the mean of the two middle values. The median separates the data into two parts such that 50% of the data values are less than the median and 50% are greater than the median.

Jonathan and Tristen are two students in Dr. Hawkes Math 230 class. So far, Dr. Hawkes has given 5 tests and the students' scores are listed below.

Jonathan	Tristen
24	80
98	84
86	88
96	72
96	81

1. Calculate the mean and median of Jonathan's grades.

2. Calculate the mean and median of Tristen's grades.

3. Compare the two measures of *average* for each student.
   a. Are the mean and median similar for Jonathan?
   b. Are the mean and median similar for Tristen?
   c. Based on the **mean**, who has the best *average* in the class?
   d. Based on the **median**, who has the best *average* in the class?

4. In your opinion, which student has the most consistent test scores? Explain your reasoning.

5. If each student had scored 2 points higher on each test, how would this affect
   a. The mean of their grades?
   b. The median?

6. Dr. Hawkes is planning on giving one more test in the class. His grading scale is as follows.

A	93–100
B	85–92
C	74–84
D	69–73
F	Below 69

   a. What is the lowest score each student can make on the test and still end up with a grade of C for the class (based on the **mean** of all test scores)?
   b. Who has to make the higher grade on the last test to get a C, Jonathan or Tristen?
   c. If the last test counts double (equivalent to two test grades) what is the lowest score each student can make on the test in order to make a B in the class (based on the **mean** of all test scores)? (Do not round the mean.)
   d. If the last test counts double, who has to make the higher grade on the last test to get a B, Jonathan or Tristen?

7. Based on the work you have done in questions 1 through 6, which measure do you think is the *best* measure of a student's *average* grade, the mean or the median? (Explain your reasoning by looking at this question from both Jonathan and Tristen's point of view.)

Name: Date: **235**

Chapter 7 Project

Misleading Graphs
An activity to demonstrate the use of bar graphs in real life.

Unfortunately, we can be intentionally or unintentionally misled by statistics, particularly when graphs are used to convey findings. In this project, you'll learn how to spot when this occurs with a bar graph and how to fix the representation.

Suppose the following table reported education levels among all young adults (18–24 years old) within the United States for a specific year.

Highest Education Level Attained	Percent (%)
Neither a High School Diploma nor a G.E.D.	5
High School Diploma or G.E.D.	65
Bachelor's Degree	20
Beyond Bachelor's Degree	10

1. The percent value in the first category is lower than all the others. For the other three categories, calculate how many times bigger the percents are compared to the percent value in the first category.

2. Suppose a graphic designer presents the information in the following bar graph. For each of the four categories, calculate the area of the bar shown. (Recall that the area of a square is $A = s^2$.)

Highest Education Level Attained

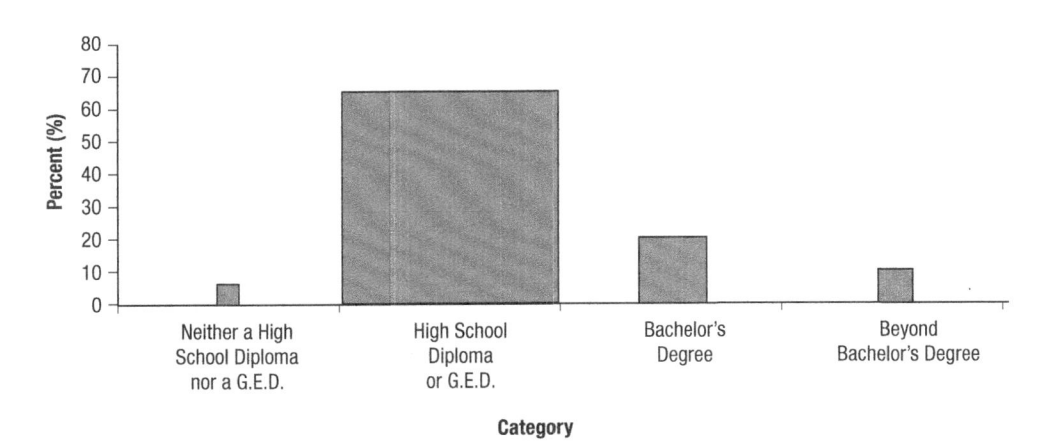

3. Similar to Problem 1, the area in the first category is smaller than the areas for all the other categories. Calculate how many times bigger the area of each of the other three are compared to the area in the first category.

4. There is a relationship between the area comparisons in Problem 3 to the percent comparisons in Problem 1. What is it? (**Hint:** What type of geometric figures are shown in the graph and what calculation does this suggest?)

5. Review Section 7.3. Which step of constructing a vertical bar graph was skipped, whether intentional or not? How could this alter the perceptions of those reading the graph in Problem 2?

6. Construct a vertical bar graph that follows all the steps shown in Section 7.3.

7. Although we are able to do the calculations and make the comparisons to spot the misrepresentation, why do you think the skipped step mentioned in Problem 5 is important? (**Note:** Whether for short reading or a deep dive, look into the works of psychologist Jean Piaget.)

8. Suppose that one year later the information was updated and presented in the following table. While it may be understandable that the information was presented this way, it cannot be used to construct a bar graph for educational levels of young adults in the US. Why?

Educational Level	Percent (%)
At Least a high School Diploma or G.E.D.	96
At Least a Bachelor's Degree	24
Beyond a Bachelor's Degree	11

9. Use the information from the table in Problem 8 to do the following.

    a. Construct a new table like the one from the beginning of the project. (**Hint:** Subtraction is required.)

    b. Explain why subtraction was required to construct the table in part a.

    c. Construct a vertical bar graph to go along with the table in part a.

CHAPTER 8

Introduction to Algebra

8.1 The Real Number Line and Absolute Value

8.2 Addition with Real Numbers

8.3 Subtraction with Real Numbers

8.4 Multiplication and Division with Real Numbers

8.5 Order of Operations with Real Numbers

8.6 Properties of Real Numbers

8.7 Simplifying and Evaluating Algebraic Expressions

8.8 Translating English Phrases and Algebraic Expressions

CHAPTER 8 PROJECTS
Going to Extremes!

Ordering Operations

Connections

The Swiss psychologist Jean Piaget once said,

"Logic and mathematics are nothing other than specialized linguistic structures."

Variables are used as place holders for numerical values in algebraic expressions and equations. Expressions and equations can be seen as tools that allow us to record, reuse, and communicate important relationships using a universally understood language.

For example, in the United States, we commonly measure temperature in degrees Fahrenheit (°F). In this temperature scale, 32 °F is the freezing temperature of water while 212 °F is the boiling point. Most of the rest of the world uses the Celsius scale (°C), where water freezes at 0 °C and boils at 100 °C. The two scales are related by the following formula.

$$C = \frac{5}{9}(F - 32)$$

According to the National Oceanic and Atmospheric Administration, the ocean water temperature at Conimicut Lighthouse in Rhode Island was 39.9 °F on February 2, 2020. How can this formula be used to determine the water temperature in degree Celsius?

8.1 The Real Number Line and Absolute Value

> **Integers**
>
> The set of numbers consisting of the _____
>
> _____
>
> **DEFINITION**

> **Variables**
>
> A **variable** is a symbol (generally a _____) that is used to _____
>
> _____
>
> **DEFINITION**

> **Rational Numbers**
>
> A **rational number** is a number that can be written in _____
>
> _____
>
> OR
>
> A **rational number** is a number that can be written in _____
>
> _____
>
> **DEFINITION**

1. Numbers that cannot be written as fractions with integer numerators and denominators are called _____ _____

▶ Watch and Work

Watch the video for Example 4 in the software and follow along in the space provided.

Example 4 Graphing Sets of Numbers

Graph the set of **real numbers** $\left\{-\frac{3}{4}, 0, 1, 1.5, 3\right\}$.

Solution

240   8.1   Exercises

✏️ Now You Try It!

Use the space provided to work out the solution to the next example.

Example A Graphing Sets of Numbers

Graph the set of real numbers $\left\{-2.5, -1, 0, \dfrac{5}{4}, 4\right\}$.

Symbols of Equality and Inequality

Reading from left to right:

= _____   ≠ _____

< _____   > _____

≤ _____   ≥ _____

Absolute Value

The **absolute value** of a real number is _____ Note that the absolute value of a _____

PROPERTIES

8.1 Exercises

Concept Check

True/False. Determine whether each statement is true or false. If a statement is false, explain how it can be changed so the statement will be true. (**Note:** There may be more than one acceptable change.)

1. On a number line, smaller numbers are always to the left of larger numbers.

2. The absolute value of a negative number is a positive number.

3. All whole numbers are also integers.

4. Zero is a positive number.

Practice

Graph each set of real numbers on a real number line.

5. $\{-3, -2, 0, 1\}$

6. $\left\{-2, -1, -\dfrac{1}{3}, 2\right\}$

List the numbers in the set $A = \left\{-7, -\sqrt{6}, -2, -\dfrac{5}{3}, -1.4, 0, \dfrac{3}{5}, \sqrt{5}, \sqrt{11}, 4, 5.9, 8\right\}$ that are described in each exercise.

7. Whole numbers

8. Rational numbers

Determine whether each statement is true or false. If a statement is false, rewrite it in a form that is a true statement. (There may be more than one way to correct a statement.)

9. $0 = -0$

10. $|-8| \geq 4$

Applications

Solve. Represent each quantity with a signed integer.

11. The Alvin is a manned deep-ocean research submersible that has explored the wreck of the Titanic. The operating depth of the Alvin is 4500 meters below sea level.

12. The Mariana trench is the deepest known location on the Earth's ocean floor. The deepest known part of the Mariana Trench is approximately 11 kilometers below sea level.

Writing & Thinking

13. Explain, in your own words, how an expression such as $-y$ might represent a positive number.

14. Compare and contrast absolute value with opposites.

8.2 Addition with Real Numbers

Rules for Addition with Real Numbers

1. To add two real numbers with **like signs**,

    a. _____

    b. use the _____

2. To add two real numbers with **unlike signs**,

    a. _____

    b. use the _____

PROCEDURE

▶ Watch and Work

Watch the video for Example 3 in the software and follow along in the space provided.

Example 3 Adding Three or More Real Numbers

Add.

a. $-3+2+(-5)$

b. $6.0+(-4.3)+(-1.5)$

Solution

✏ Now You Try It!

Use the space provided to work out the solution to the next example.

Example A Adding Three or More Real Numbers

Add.

a. $-7+5+(-3)$

b. $-3.2+(-6.1)+5.7$

Solution

8.2 Exercises

Concept Check

True/False. Determine whether each statement is true or false. If a statement is false, explain how it can be changed so the statement will be true. (**Note:** There may be more than one acceptable change.)

1. The sum of a positive number and a negative number is always positive.

2. When adding two numbers with unlike signs, the result uses the sign of the number with the larger absolute value.

3. The sum of two positive numbers can equal zero.

Practice

Add. Reduce any fractions to lowest terms.

4. $8 + (-3)$

5. $2 + (-8)$

6. $-\dfrac{1}{6} + \dfrac{7}{15}$

7. $3.2 + (-1.2) + (-2.5)$

Add. Be sure to find the absolute values first.

8. $13 + |-5|$

Applications

Solve.

9. For 2024, a business reports a profit of $45,000 during the first quarter, a loss of $8000 during the second quarter, a loss of $2000 during the third quarter, and a profit of $15,000 during the fourth quarter.

    a. Write an addition expression to represent the total profit made by the company in 2024. Do not simplify.

    b. Simplify the expression from part a.

10. A submarine dives to a depth of 250 feet below the surface. It rises 75 feet before diving an additional 100 feet. What is the final depth of the submarine?

Writing & Thinking

11. Describe, in your own words, how the sum of the absolute values of two numbers might be 0. (Is this even possible?)

12. Describe in your own words the conditions under which the sum of two integers will be 0.

Name: _____   Date: _____   **247**

8.3 Subtraction with Real Numbers

> **Additive Inverse**
>
> The opposite of a _____. The sum of a number and its additive
>
> inverse _____. Symbolically, for any real number a,
>
> _____
>
> **DEFINITION**

> **Subtraction**
>
> For any real numbers a and b,
>
> _____
>
> In words, to subtract b from a, _____
>
> **DEFINITION**

1. To find the **change in value** between two numbers, _____

   _____. Symbolically,

   change in value = _____.

▶ Watch and Work

Watch the video for Example 4 in the software and follow along in the space provided.

Example 4 Application: Calculating Change in Value

A jet pilot flew her plane from an altitude of 30,000 ft to an altitude of 12,000 ft. What was the change in altitude?

Solution

✏ Now You Try It!

Use the space provided to work out the solution to the next example.

Example A Application: Calculating Change in Value

A drone plane flew from an altitude of 25,000 ft to an altitude of 14,000 ft. What was the change in altitude?

8.3 Exercises

Concept Check

True/False. Determine whether each statement is true or false. If a statement is false, explain how it can be changed so the statement will be true. (**Note:** There may be more than one acceptable change.)

1. The sum of a number and its additive inverse is the number itself.

2. The additive inverse of negative seven is seven.

3. We can think of addition of numbers as accumulating numbers.

4. The expression "15 – 7" can be thought of as "fifteen plus negative seven."

Practice

Find the additive inverse (opposite) of each real number.

5. 11

6. −3.4

Subtract. Reduce fractions to lowest terms.

7. $-8-(-11)$

8. $\dfrac{7}{15} - \dfrac{2}{15}$

Perform the indicated operation to find the net change in value.

9. $-6+(-4)-5$

10. $-11.3 + 5.3 - 7.9$

Applications

Solve.

11. At 2 p.m. the temperature was 76 °F. At 8 p.m. the temperature was 58 °F. What was the change in temperature?

12. A couple sold their house for $175,000. They paid the realtor $9100, and other expenses of the sale came to $800. If they owed the bank $126,000 for the mortgage, what were their net proceeds from the sale?

Writing & Thinking

13. Explain, in your own words, how to find the difference between a positive and a negative number.

14. What is the additive inverse of 0? Why?

Name: Date: **251**

8.4 Multiplication and Division with Real Numbers

Rules for Multiplication with Real Numbers

If a and b are positive real numbers, then

1. The product of two positive numbers _____
2. The product of two negative numbers _____
3. The product of a positive number and a negative number _____
4. The product of 0 and any number _____

PROCEDURE

Division with Real Numbers

For real numbers a, b, and x (where $b \neq 0$),

$$\frac{a}{b} = x \text{ means } \underline{\hspace{2cm}}$$

For real numbers a and b (where $b \neq 0$),

$$\frac{a}{0} \text{ is } \underline{\hspace{2cm}}$$

DEFINITION

Rules for Division with Real Numbers

If a and b are positive real numbers (where $b \neq 0$),

1. The quotient of two positive numbers _____
2. The quotient of two negative numbers _____
3. The quotient of a positive number and a negative number _____

PROCEDURE

Average

The **average** (or **mean**) of a set of numbers is the value found by _____
_____

DEFINITION

8.4 Multiplication and Division with Real Numbers

▶ Watch and Work

Watch the video for Example 6 in the software and follow along in the space provided.

Example 6 Application: Calculating an Average

At noon on five consecutive days in Aspen, Colorado, the temperatures were −5°, 7°, 6°, −7°, and 14° (in degrees Fahrenheit). (Negative numbers represent temperatures below zero). Find the average of these noonday temperatures.

Solution

✎ Now You Try It!

Use the space provided to work out the solution to the next example.

Example A Application: Calculating an Average

At noon on five consecutive days in Mears, Michigan, the temperatures were −3°, 5°, 8°, −4°, and 14° (in degrees Fahrenheit.) (Negative numbers represent temperatures below zero.) Find the average of these noonday temperatures.

8.4 Exercises

Concept Check

True/False. Determine whether each statement is true or false. If a statement is false, explain how it can be changed so the statement will be true. (**Note:** There may be more than one acceptable change.)

1. If a negative number is divided by a positive number, the result will be a negative number.

2. The product of zero and a number is zero.

3. If two numbers have the same sign, both the product and the quotient of the two numbers will be negative.

4. The mean of a set of numbers is always positive.

Practice

Multiply. Reduce fractions to lowest terms.

5. $12 \cdot 4$

6. $(-7)(-16)(0)$

Divide. Reduce fractions to lowest terms. Round answers with decimals to the nearest tenth.

7. $\dfrac{-20}{-10}$

8. $\dfrac{-5.6}{7}$

Applications

Solve.

9. Find the mean of the following set of integers: $-10, 15, 16, -17, -34,$ and -42.

10. According to the US Fish and Wildlife Service, migratory birds are imported at a value of about $19 each. Suppose that about 800,000 live birds are imported each year. What is the total value of these imported birds?

Writing & Thinking

11. If you multiply an odd number of negative numbers together, do you think that the product will be positive or negative? Explain your reasoning.

12. Explain the conditions under which the quotient of two numbers is 0.

8.5 Order of Operations with Real Numbers

Rules for Order of Operations
1. Simplify within grouping symbols, such as _____
   _____
2. Find any _____
3. Moving from left to right, perform any _____
4. Moving from left to right, perform any _____

PROCEDURE

▶ Watch and Work

Watch the video for Example 3 in the software and follow along in the space provided.

Example 3 Using the Order of Operations with Real Numbers

Simplify: $(2-5)^2 + |2-5^2| - 2^3$

Solution

✏ Now You Try It!

Use the space provided to work out the solution to the next example.

Example A Using the Order of Operations with Real Numbers

Simplify: $(1-3)^2 + |9-4^2| - 1^3$

8.5 Exercises

Concept Check

True/False. Determine whether each statement is true or false. If a statement is false, explain how it can be changed so the statement will be true. (**Note:** There may be more than one acceptable change.)

1. If there are no grouping symbols, multiplication should always be performed before addition.

2. When following the rules for order of operations, powers indicated by exponents should be evaluated last.

3. The square root symbol is a grouping symbol.

4. A well-known mnemonic device for remembering the rules for order of operations is SADMEP.

Practice

Simplify.

5. a. $24 \div 4 \cdot 6$

    b. $24 \cdot 4 \div 6$

6. $15 \div (-3) \cdot 3 - 10$

7. $3^2 \div (-9) \cdot (4 - 2^2) + 5(-2)$

8. $14 \cdot 3 \div (-2) - 6(4)$

9. $|16 - 20| + (-10)^2 + 5^2$

Applications

Solve.

10. The Matthews family, a family of 4, is planning a trip to New York City. During their visit, they want to see the Broadway play *Beetlejuice*. The tickets cost $102 each. The Matthews purchase the tickets online and the website charges a service fee of $7.50 per ticket. The website is running a sale where the Matthews can get 10% off of their entire purchase.

   a. Write an expression to describe how much of a discount the Matthews will receive on their purchase.

   b. What is the final purchase price of the tickets?

11. Dennis overdrew his checking account and ended up with a balance of −$42. The bank charged a $35 overdraft fee and an additional $5 fee for every day the account was overdrawn. Dennis left his account overdrawn for 3 days.

   a. Write an expression to show the balance of Dennis's checking account after 3 days.

   b. Simplify the expression in part a. to find the balance of Dennis's checking account after 3 days.

Writing & Thinking

12. Explain, in your own words, why the following expression cannot be evaluated.

$$\left(24-2^4\right)+6(3-5)\div\left(3^2-9\right)$$

13. Consider any number between 0 and 1. If you square this number, will the result be larger or smaller than the original number? Is this always the case? Explain.

8.6 Properties of Real Numbers

Properties of Addition and Multiplication

In this table a, b, and c are real numbers.

Name of Property	For Addition	For Multiplication
Commutative property	_____	_____
	_____	_____
Associative property	_____	_____
	_____	_____
Identity	_____	_____
	_____	_____
Inverse	_____	_____
	_____	_____

Zero-Factor Law

_____   _____

Distributive Property of Multiplication over Addition

_____   _____

PROPERTIES

▶ Watch and Work

Watch the video for Example 2 in the software and follow along in the space provided.

Example 2 Identifying Properties of Addition and Multiplication

For each of the following equations, state the property illustrated, and show that the statement is true for the value given for the variable by substituting the value in the equation and evaluating.

a. $x + 14 = 14 + x$ given that $x = -4$

b. $(3 \cdot 6)x = 3(6x)$ given that $x = 5$

c. $12(y + 3) = 12y + 36$ given that $y = -2$

Solution

✏ Now You Try It!

Use the space provided to work out the solution to the next example.

Example A Identifying Properties of Addition and Multiplication

State the property illustrated and show that the statement is true for the value given for the variable.

a. $x + 21 = 21 + x$ given that $x = -7$

b. $(5 \cdot 4)x = 5(4x)$ given that $x = 2$

c. $11(y + 3) = 11y + 33$ given that $y = -4$

8.6 Exercises

Concept Check

True/False. Determine whether each statement is true or false. If a statement is false, explain how it can be changed so the statement will be true. (**Note:** There may be more than one acceptable change.)

1. Changing the order of the numbers in an addition problem is allowed because of the associative property of addition.

2. The equation $(8 \cdot 2) \cdot 5 = 8 \cdot (2 \cdot 5)$ is an example of the associative property of multiplication.

3. The additive identity of all numbers is 1.

4. The commutative property works for division and subtraction.

Practice

Complete the expressions using the given property. Do not simplify.

5. $7 + 3 =$ _____ commutative property of addition

6. $(6 \cdot 9) \cdot 3 =$ _____ associative property of multiplication

7. $19 \cdot 4 =$ _____ commutative property of multiplication

8. $6(5 + 8) =$ _____ distributive property

9. $16 + (9 + 11) =$ _____ associative property of addition

10. $6 \cdot 0 =$ _____ zero-factor law

Applications

Solve

11. Jessica works part-time at a retail store and makes $11 an hour. During one week, she worked $6\frac{1}{2}$ hours on Monday and $4\frac{1}{4}$ hours on Thursday.

    a. Determine the amount of money she earned during the week by evaluating the expression $\$11 \cdot \left(6\frac{1}{2} + 4\frac{1}{4}\right)$.

    b. Rewrite this expression to remove the parentheses using one of the properties talked about in this section.

    c. What property did you use in part b. to rewrite the expression?

12. Robin went to the grocery store to buy a few items she needed in order to cook dinner. She bought milk for $3.99, rolls for $2.25, a package of steaks for $12.01, and some marinade for $1.75. Before getting to the checkout line, Robin remembered that she only had $20 in her purse. Did she have enough money to buy the food items if the store does not charge sales tax on food?

    a. Write an expression to find the total of Robin's food purchases. Do not simplify.

    b. Robin doesn't have a calculator to determine the total cost of her items. She wants to make sure that she has enough money to buy them. Rearrange the expression from part a. so that she could quickly find the total using mental math.

    c. What properties did you use in part b. to rewrite the expression?

    d. Did Robin have enough money to purchase all of the items?

Writing & Thinking

13. a. The distributive property illustrated as $a(b+c) = ab + ac$ is said to "distribute multiplication over addition." Explain, in your own words, the meaning of this phrase.

    b. What would an expression that "distributes addition over multiplication" look like? Explain why this would or would not make sense.

Name: _____ Date: _____ **263**

8.7 Simplifying and Evaluating Algebraic Expressions

Like Terms

Like terms (or similar terms) are terms that are _____
_____

DEFINITION

Combining Like Terms

To combine like terms, _____

DEFINITION

Evaluating an Algebraic Expression

1. _____
2. _____
3. _____

(**Note:** Terms separated by + and − signs may be evaluated _____
_____.)

PROCEDURE

▶ Watch and Work

Watch the video for Example 5 in the software and follow along in the space provided.

Example 5 Simplifying and Evaluating Algebraic Expressions

Simplify and evaluate $3ab - 4ab + 6a - a$ for $a = 2$, $b = -1$.

Solution

Now You Try It!

Use the space provided to work out the solution to the next example.

Example A Simplifying and Evaluating Algebraic Expressions

Simplify and evaluate $5ab - 8ab + 2a - 3a$ for $a = -3$, $b = 1$.

8.7 Exercises

Concept Check

True/False. Determine whether each statement is true or false. If a statement is false, explain how it can be changed so the statement will be true. (**Note:** There may be more than one acceptable change.)

1. A variable that does not appear to have an exponent has an exponent of 1.

2. In the term $-9x$, nine is being subtracted from x.

3. In the term "$12a$," 12 is the constant.

4. Like terms have the same coefficients.

Practice

Identify the like terms in each list of terms.

5. $-5, 3, 7x, 8, 9x, 3y$

Simplify each expression by combining like terms.

6. $8x + 7x$

7. $3x - 5x + 12x$

8. $13x + 12x^2 + 15x - 35 - 41 - 2x^2$

Simplify each expression and then evaluate the expression for $y = 3$ and $a = -2$.

9. $5y + 4 - 2y$

10. $\dfrac{3a + 5a}{-2} + 12a$

Applications

Solve.

11. An apartment management company owns a property with 100 units. The company has determined that the profit made per month from the property can be calculated using the equation $P = -10x^2 + 1500x - 6000$, where x is the number of units rented per month. How much profit does the company make when 80 units are rented?

12. A ball is thrown upward from an initial height of 96 feet with an initial velocity of 16 feet per second. After t seconds, the height of the ball can be described by the expression $-16t^2 + 16t + 96$. What is the height of the ball after 3 seconds?

Writing & Thinking

13. Discuss like and unlike terms and give an example of each.

14. Explain the difference between -13^2 and $(-13)^2$.

8.8 Translating English Phrases and Algebraic Expressions

Key Words To Look For When Translating Phrases

Addition	Subtraction	Multiplication	Division	Exponent (Powers)

1. Division and subtraction are done with the values in the _____ that they are given in the problem.

2. An **ambiguous phrase** is one whose meaning is _____
   _____

▶ Watch and Work

Watch the video for Example 3 in the software and follow along in the space provided.

Example 3 Translating Algebraic Expressions into English Phrases

Change each algebraic expression into an equivalent English phrase. In each case translate the variable as "a number."

a. $5x$

b. $2n + 8$

c. $3(a-2)$

Solution

✏️ Now You Try It!

Use the space provided to work out the solution to the next example.

Example A Translating Algebraic Expressions into English Phrases

Change each algebraic expression into an equivalent English phrase.

a. $10x$

b. $4a + 7$

c. $7(n-5)$

8.8 Exercises

Concept Check

True/False. Determine whether each statement is true or false. If a statement is false, explain how it can be changed so the statement will be true. (**Note:** There may be more than one acceptable change.)

1. The order in which the values are given is particularly important when working with subtraction and division problems.

2. "More than" and "increased by" are key phrases specifying the operation of subtraction.

3. Division is indicated by the phrase "five less than a number."

4. Key phrases for parentheses can be used to limit ambiguity in English phrases.

Practice

Write the algebraic expressions described by the English phrases. Choose your own variable.

5. six added to a number

6. twenty decreased by the product of four and a number

7. eighteen less than the quotient of a number and two

Translate each pair of English phrases into algebraic expressions. Notice the differences between the algebraic expressions and the corresponding English phrases.

8. **a.** six less than a number

   b. six less a number

9. **a.** six less than four times a number

   b. six less four times a number

Write the algebraic expression described by the English phrase using the given variables.

10. the cost of purchasing a fishing rod and reel if the rod costs x dollars and the reel costs $8 more than twice the cost of the rod

Translate each algebraic expression into an equivalent English phrase. (There may be more than one correct translation.)

11. $-9x$

12. $\dfrac{9}{x+3}$

Writing & Thinking

13. Explain why translating addition and multiplication problems from English into algebra may be easier than changing subtraction or division problems. (Consider the properties previously studied.)

14. Explain the difference between $5(n+3)$ and $5n+3$ when converting from algebra to English.

Chapter 8 Project

Going to Extremes!

An activity to demonstrate the use of signed numbers in real life.

When asked what the highest mountain peak in the world is, most people would say Mount Everest. This answer may be correct, depending on what you mean by highest. According to Geology.com, there may be other contenders for this important distinction.

The peak of Mount Everest is 8850 meters (or 29,035 feet) above sea level, giving it the distinction of being the mountain with the highest altitude in the world. However, Mauna Kea is a volcano on the big island of Hawaii whose peak is over 10,000 meters above the nearby ocean floor, which makes it taller than Mount Everest. A third contender for the highest mountain peak is Chimborazo, an inactive volcano in Ecuador. Although Chimborazo only has an altitude of 6263 meters (20,549 feet) above sea level, it is the highest mountain above Earth's center. How could a mountain that is only 6263 meters tall be higher than a mountain that is 8850 meters tall? Because the Earth is not really a sphere but an "oblate spheroid," which means that the Earth is widest at the equator. Chimborazo is 1° south of the equator which makes it about 2000 meters farther from the Earth's center than Mount Everest.

What about the other extreme? What is the lowest point on Earth? As you might have guessed, there is also more than one candidate for that distinction. The lowest exposed area of land on Earth's surface is on the Dead Sea shore at 413 meters below sea level. The Bentley Subglacial Trench in Antarctica is the lowest point on Earth that is not covered by ocean but is covered by ice. This trench reaches 2555 meters below sea level. The deepest point on the ocean floor occurs 10,916 meters below sea level in the Mariana Trench in the Pacific Ocean.

1. Calculate the **difference** in elevation between Mount Everest and Chimborazo in both meters and feet. What operation does the word **difference** imply?

2. Write an expression to calculate the **difference** in elevation between the peak of Mount Everest and the lowest point on the Dead Sea shore in meters and then simplify.

3. If you were to travel from the bottom of the Mariana Trench to the top of Mount Everest, how many meters would you travel?

4. If Mount Everest were magically moved and placed at the bottom of the Mariana Trench, how many meters of water would lie above Mount Everest's peak?

5. How much farther below sea level (in meters) is the Mariana Trench as compared to the Dead Sea shore?

6. Add the elevations (in meters) together for Mount Everest, Chimborazo, the Dead Sea Shore, the Bentley Subglacial Trench, and the Mariana Trench and show your result. Is this number positive or negative? Would this value represent an elevation above or below sea level?

7. Convert the results in Problems 2 through 4 above to feet using the conversion factor 1 meter = 3.28 feet. Do not round your answers.

8. Convert the results in Problem 7 from feet to miles using the conversion factor 1 mile = 5280 feet. (Round your answers to the nearest thousandth.)

9. Using the height of Mount Everest in meters as an example, the conversions in Problems 7 and 8 could have been combined to do the conversion from meters directly to miles by using the following sequence of conversion factors:

$$8850 \text{ m} \cdot \frac{3.25 \text{ ft}}{1 \text{ m}} \cdot \frac{1 \text{ mi}}{5280 \text{ ft}} \approx 5.498 \text{ miles}$$

(A mountain peak over 5 miles high!)

Now notice that in doing the conversions, the units for meters and feet cancel out since they appear in both the numerator and denominator, leaving only the unit of miles in the numerator of your result. This is called dimensional analysis and is extremely helpful in converting measurements to make sure you end up with the correct answer and the correct units on your result.

Now verify that this sequence of conversions works by taking the results from Problems 2 through 4 and applying both conversion factors above. How do your results compare to the results from Problem 8? (Round your answers to the nearest thousandth.)

10. There is more than one way that this conversion could have been performed. Using the conversion factors 1 km = 1000 m and 1 mile = 1.61 km, convert the results in Problems 2 through 4 from meters to miles by using these factors in sequence similar to Problem 9 and performing a dimensional analysis. (Round your answer to the nearest thousandth.) Do you get exactly the same results? Why do you think this is so?

Chapter 8 Project

Ordering Operations
An activity to explore the orders of operations.

Did you know that the order of operations you learn in math courses is a relatively new set of rules? The rules were created in the early 1900s and solidified into the current form along with the creation of computers and computer languages. Before the 1600s, mathematical notation was not commonly used, and mathematical expressions and equations were written out in words. Any phrasing that was ambiguous (that is, could be understood in more than one way) was avoided. When simplifying expressions in mathematics, care must be taken to properly follow the order of operations. If you stray from the order, you will reach a different conclusion than intended.

Suppose you are creating a computer program to follow the rules for the order of operations. The computer uses * for multiplication, / for division, and ^ to indicate exponents.

1. As your first test, you enter "8 + 0 * 3 − 10 / 2" into the computer program and it returns a value of 7.
   a. Following the order of operations, what would you expect as the result?
   b. Explain the error(s) made by the computer program.
   c. Insert grouping symbols into the expression so that it will simplify to the value returned by the computer.
   d. Rewrite the expression (using parentheses if needed) so that the computer program will return the expected value from part a. if the computer program is not modified.
   e. Compare the answers from parts c. and d. What do you notice?

2. After modifying your code, you verify that the previous expression returns the correct value. For the next test, you enter "−4^2 − 10 / 2 + 4" and the computer returns a value of 15.
   a. Following the order of operations, what would you expect as the result?
   b. Explain the error(s) made by the computer program.
   c. Rewrite the original expression so that it will simplify to the value returned by the computer.

3. After further modification, the computer program can now properly follow the order of operations. You next add code to allow the program to simplify algebraic expressions. You test your code by entering "−2(3x + 5y)" and it returns −16xy.
   a. What would you expect the program to return?
   b. Explain the error(s) made by the computer program.
   c. Is this the same error as the computer returning a result of 30xy when you enter "3(4x + 6y)"?

4. As a final step, you write code to allow the program to translate English phrases into algebraic expressions. You enter the phrase "five times two plus three" and the program returns "5 * 2 + 3".
   a. If you expected "5(2 + 3)" in return, is the issue with the computer program or your phrasing? Explain what the issue is.
   b. Next, you enter the phrase "twelve less two times a number" and the program returns "2x − 12". Is the issue the computer program or phrasing?

CHAPTER 9

Solving Linear Equations and Inequalities

9.1 Solving Linear Equations: $x + b = c$ and $ax = c$

9.2 Solving Linear Equations: $ax + b = c$

9.3 Solving Linear Equations: $ax + b = cx + d$

9.4 Working with Formulas

9.5 Applications: Number Problems and Consecutive Integers

9.6 Applications: Distance-Rate-Time, Interest, Average, and Cost

9.7 Solving Linear Inequalities in One Variable

9.8 Compound Inequalities

9.9 Absolute Value Equations

9.10 Absolute Value Inequalities

CHAPTER 9 PROJECTS
A Linear Vacation

Breaking Even

Connections

You have probably heard that mathematics is the language of science, engineering, and business. While that is considered to be true, it can be difficult to see where some of the skills you develop in a math course are applied in those areas.

For example, the ability to solve linear equations is particularly useful in the study of free markets in economics. Economists want to be able to understand how changes in the price of goods affect consumer behavior. One way to gain this understanding is to analyze the demand for a specific product.

The demand of a product is the number of units q that the market is willing to absorb at a price p. In other words, it's how many items consumers are willing and able to purchase. Typically, the lower the price of a product, the more units consumers will be willing to purchase. To illustrate this relationship, consider the US demand for wheat in 1981, which can be illustrated by the following equation.

$$q = 3550 - 266p$$

In this equation, q is measured in millions of bushels and p is the price per bushel in dollars.

How can you determine the price per bushel when the demand hit 2630 million bushels?

Name: _____ Date: _____ **277**

9.1 Solving Linear Equations: $x + b = c$ and $ax = c$

Linear Equation in x

If a, b, and c are **constants** and $a \neq 0$ then a **linear equation in x** is _____

_____

_____

DEFINITION

Addition Principle of Equality

If the same algebraic expression is added to _____

_____

Symbolically, if A, B, and C are algebraic expressions, then the equations

_____

and

_____

_____

PROPERTIES

Solving Linear Equations that Simplify to the Form $x + b = c$

1. Combine _____

2. Use the addition principle of equality and _____

   The objective is to _____

   _____

3. Check your answer by _____

PROCEDURE

▶ Watch and Work

Watch the video for Example 2 in the software and follow along in the space provided.

Example 2 Solving Linear Equations of the Form $x + b = c$

Solve the equation: $x - 3 = 7$

9.1 Solving Linear Equations: $x + b = c$ and $ax = c$

Solution

✏️ Now You Try It!

Use the space provided to work out the solution to the next example.

Example A Solving Linear Equations of the Form $x + b = c$

Solve the equation: $x - 5 = 12$

Multiplication (or Division) Principle of Equality

If both sides of an equation are multiplied by (or divided by) the _____
_____

Symbolically, if A and B are algebraic expressions and C is any nonzero constant, then the equations

_____

_____

_____

_____

PROPERTIES

> **Solving Linear Equations that Simplify to the Form** $ax = c$
> 1. Combine _____
>
> 2. Use the **multiplication** (or **division**) **principle of equality** and multiply both sides of the equation by
>
>   the _____ (or divide both sides by _____
>
>   _____). The coefficient of the variable will become _____
>
> 3. Check your answer by _____
>
> PROCEDURE

9.1 Exercises

Concept Check

True/False. Determine whether each statement is true or false. If a statement is false, explain how it can be changed so the statement will be true. (**Note:** There may be more than one acceptable change.)

1. When an algebraic expression is added to both sides of an equation, the new equation has the same solutions as the original equation.

2. The process of finding the solution set to an equation is called simplifying the equation.

3. A linear equation in x is also called a first-degree equation in x.

4. Equations with the same solutions are said to be equivalent equations.

Practice

Determine whether the given number is a solution to the given equation by substituting and then evaluating.

5. $x - 2 = -3$ given that $x = 1$

6. $-10 + x = -14$ given that $x = -4$

9.1 Exercises

Solve each equation.

7. $x - 6 = 1$

8. $x - 10 = 9$

9. $51 = 17y$

10. $\dfrac{5x}{7} = 65$

Applications

Solve.

11. The Japanese writing system consists of three sets of characters, two with 81 characters (which all Japanese students must know), and a third, *kanji*, with over 50,000 characters (of which only some are used in everyday writing). If a Japanese student knows 2107 total characters, solve the equation $x + 2(81) = 2107$ to determine the number of *kanji* characters the student knows.

12. A nurse must give a patient 800 milliliters of intravenous solution over 4 hours. This can be represented by the equation $4x = 800$, where x represents the amount of solution the patient receives per hour in milliliters.

    a. Why was multiplication chosen in the equation?

    b. Solve the equation to determine the value of x.

    c. What does the answer to part b. mean? Write a complete sentence.

Writing & Thinking

13. a. Is the expression $6 + 3 = 9$ an equation? Explain.

    b. Is 4 a solution to the equation $5 + x = 10$? Explain.

9.2 Solving Linear Equations: *ax + b = c*

Solving Linear Equations that Simplify to the Form *ax + b = c*

1. Combine _____

2. Use the **addition principle of equality** and _____
   _____

3. Use the _____ and multiply both sides of the equation by the reciprocal of the coefficient of the variable (or _____
   _____ itself). The _____ will become +1.

4. Check your answer by _____

PROCEDURE

▶ Watch and Work

Watch the video for Example 2 in the software and follow along in the space provided.

Example 2 Solving Linear Equations of the Form *ax + b = c*

Solve the equation: $-26 = 2y - 14 - 4y$

Solution

Now You Try It!

Use the space provided to work out the solution to the next example.

Example A Solving Linear Equations of the Form $ax + b = c$

Solve the equation.

$-18 = 2y - 8 - 7y$

9.2 Exercises

Concept Check

True/False. Determine whether each statement is true or false. If a statement is false, explain how it can be changed so the statement will be true. (**Note:** There may be more than one acceptable change.)

1. If an equation of the form $ax + b = c$ uses decimal or fractional coefficients, the addition and multiplication principles of equality cannot be used.

2. The first step in solving $2x + 3 = 9$ is to add 3 to both sides.

3. To solve an equation that has been simplified to $4x = 12$, you need to multiply both sides by $\frac{1}{4}$, or divide both sides by 4.

4. When solving a linear equation with decimal coefficients, one approach is to multiply both sides in such a way to give integer coefficients before solving.

Practice

Solve each equation.

5. $3x + 11 = 2$

6. $-5x + 2.9 = 3.5$

7. $\dfrac{2}{5} - \dfrac{1}{2}x = \dfrac{7}{4}$

8. $\dfrac{y}{3} - \dfrac{2}{3} = 7$

Applications

Solve.

9. The tickets for a concert featuring the new hit band, Flying Sailor, sold out in 2.5 hours. If there were 35,000 tickets sold, solve the equation $35{,}000 - 2.5x = 0$ to find the number of tickets sold per hour.

10. All snacks (candy, popcorn, and soda) cost $3.50 each at the local movie theater. Admission tickets cost $7.50 each. After a long week, Carlos treats himself to a night at the movies. His movie night budget is $25 and he spends all his movie money. Solve the equation $3.50x + $7.50 = $25.00 to determine how many snacks Carlos can buy.

Writing & Thinking

11. Find the error(s) made in solving each equation and give the correct solution.

   a.
   $$\frac{1}{3}x + 4 = 9$$
   $$3 \cdot \frac{1}{3}x + 4 = 3 \cdot 9$$
   $$x + 4 = 27$$
   $$x + 4 - 4 = 27 - 4$$
   $$x = 23$$

   b.
   $$5x + 3 = 11$$
   $$(5x - 3) + (3 - 3) = 11 - 3$$
   $$2x + 0 = 8$$
   $$\frac{2x}{2} = \frac{8}{2}$$
   $$x = 4$$

9.3 Solving Linear Equations: $ax + b = cx + d$

Solving Linear Equations that Simplify to the Form $ax + b = cx + d$

1. Simplify each side of the equation by _____
   _____

2. Use the **addition principle of equality** and _____
   _____

3. Use the **multiplication** (or **division**) **principle of equality** and _____
   _____ of the coefficient of the variable (**or divide** _____ itself).
   The coefficient of the variable _____

4. Check your answer by _____

PROCEDURE

Type of Equation	Number of Solutions
conditional	_____
identity	_____
contradiction	_____

Table 1

▶ Watch and Work

Watch the video for Example 9 in the software and follow along in the space provided.

Example 9 Determining Types of Equations

Determine whether the equation $3(x-25)+3x = 6(x+10)$ is a conditional equation, an identity, or a contradiction.

Solution

Now You Try It!

Use the space provided to work out the solution to the next example.

Example A Determining Types of Equations

Determine whether the equation $-3(x - 4) = 12 - 2x - x$ is a conditional equation, an identity, or a contradiction.

9.3 Exercises

Concept Check

True/False. Determine whether each statement is true or false. If a statement is false, explain how it can be changed so the statement will be true. (**Note:** There may be more than one acceptable change.)

1. Every linear equation has exactly one solution.

2. If a linear equation simplifies to a statement that is always true, then the original equation is called an identity.

3. If an equation has no solution, it is called an identity.

4. The most general form of a linear equation is $ax + b = cx + d$.

Practice

Solve each equation.

5. $3x + 2 = x - 8$

6. $2(z+1) = 3z + 3$

7. $x - 0.1x + 0.8 = 0.2x + 0.1$

8. $0.6x - 22.9 = 1.5x - 18.4$

Determine whether each equation is a conditional equation, an identity, or a contradiction.

9. $-2x + 13 = -2(x - 7)$

10. $3x + 9 = -3(x - 3) + 6x$

9.3 Exercises

Applications

Solve.

11. Caitlyn and Steve are planning their wedding reception and must decide between two catering halls. The first site, A Wedding Space, rents for $800 for one day and charges $50 per person for dinner. The second venue, A Wedding Place, costs $1000 to rent for one day and charges $40 per person for the same dinner. Solve the equation $800 + 50x = 1000 + 40x$ to determine how many guests they can invite so that the cost they pay will be the same at both wedding catering halls.

12. The value of a new car depreciates at a rate of about $250 per month. Suppose a car originally costs $30,000. The car was bought with a $1000 down payment and a loan with 0% financing for 60 months with payments of $200 a month. Solve the equation $30{,}000 - 250t = 29{,}000 - 200t$ to determine how many months it will take for the value of the vehicle to equal the amount owed on the loan?

Writing & Thinking

13. Answer each question.

    a. Simplify the expression $3(x+5)+2(x-7)$.

    b. Solve the equation $3(x+5)+2(x-7)=31$.

    c. How are the methods you used to answer parts a. and b. similar? How are they different?

Name: _____   Date: _____   **289**

9.4 Working with Formulas

1. **Formulas** are general rules or _____ stated _____.

2. We say that the formula $d = rt$ is _____ d _____ r and t. Similarly, the formula $A = \frac{1}{2}bh$ is solved for _____ in terms of _____, and the formula $P = R - C$ (profit is equal to revenue minus cost) is solved for _____ in terms of _____.

▶ Watch and Work

Watch the video for Example 6 in the software and follow along in the space provided.

Example 6 Solving for Different Variables

Given $V = \dfrac{k}{P}$, solve for P in terms of V and k.

Solution

9.4 Exercises

✏️ Now You Try It!

Use the space provided to work out the solution to the next example.

Example A Solving for Different Variables

Given $P = \dfrac{I}{rt}$ solve for t in terms of I, r, and P.

9.4 Exercises

Concept Check

True/False. Determine whether each statement is true or false. If a statement is false, explain how it can be changed so the statement will be true. (**Note:** There may be more than one acceptable change.)

1. When using formulas, typically it does not matter if capital or lower case letters are used: $A = a$, $C = c$, etc.

2. If the perimeter and length are known, $P = 2l + 2w$ can be used to find the width of a rectangle.

3. Rate of interest is stated as an annual rate in percent form.

Applications

In the following application problems, read the descriptions carefully and then substitute the values given in the problem for the corresponding variables in the formulas. Evaluate the resulting expression for the unknown variable.

Velocity

If an object is shot upward with an initial velocity v_0 in feet per second, the velocity v in feet per second is given by the formula $v = v_0 - 32t$, where t is time in seconds. (v_0 is read "v sub zero." The $_0$ is called a subscript.)

4. An object projected upward with an initial velocity of 106 feet per second has a velocity of 42 feet per second. How many seconds have passed?

Investments

The total amount of money in an account with P dollars invested in it is given by the formula $A = P + Prt$, where r is the rate expressed as a decimal and t is time (one year or part of a year).

5. If $1000 is invested at 6% interest, find the total amount in the account after 6 months.

Cost

The total cost C of producing x items can be found by the formula $C = ax + k$, where a is the cost per item and k is the fixed costs (rent, utilities, and so on).

6. Find the total cost of producing 30 items if each costs $15 and the fixed costs are $580.

9.4 Exercises

Solve each formula for the indicated variable.

7. $P = a + b + c$; solve for b.

8. $P = 3s$; solve for s.

9. $I = Prt$; solve for t.

10. $A = P(1 + rt)$; solve for r.

9.5 Applications: Number Problems and Consecutive Integers

Consecutive Integers

Integers are **consecutive** if each is _____

Three consecutive integers can be represented as

_____

where n is an integer.

DEFINITION

Consecutive Even Integers

Even integers are **consecutive** if each is _____

Three consecutive even integers can be represented as

_____

where n is an **even** integer.

DEFINITION

Consecutive Odd Integers

Odd integers are **consecutive** if each is _____

Three consecutive odd integers can be represented as

_____

where n is an **odd** integer.

DEFINITION

▶ Watch and Work

Watch the video for Example 6 in the software and follow along in the space provided.

Example 6 Application: Calculating Living Expenses

Joe wants to budget $\frac{2}{5}$ of his monthly income for rent. He found an apartment he likes for $800 a month. What monthly income does he need to be able to afford this apartment?

Solution

✏️ Now You Try It!

Use the space provided to work out the solution to the next example.

Example A Application: Calculating Living Expenses

Jim plans to budget $\frac{3}{7}$ of his monthly income to send his son Taylor to private school. If the school he'd like Taylor to attend costs $1200 a month, what monthly income does Jim need to be able to afford the school?

9.5 Exercises

Concept Check

True/False. Determine whether each statement is true or false. If a statement is false, explain how it can be changed so the statement will be true. (**Note:** There may be more than one acceptable change.)

1. If an odd integer is divided by 2, the remainder will be 1.

2. To find 3 consecutive odd integers, you could use n, $n + 1$, and $n + 3$.

3. Odd integers are integers that are divisible by 1.

4. Even integers are consecutive if each is 2 more than the previous even integer.

Practice

Read each problem carefully, translate the various phrases into algebraic expressions, set up an equation, and solve the equation.

5. Five less than a number is equal to 13 decreased by the number. Find the number.

6. Twice a number increased by 3 times the number is equal to 4 times the sum of the number and 3. Find the number.

7. Find three consecutive integers whose sum is 93.

8. Find three consecutive odd integers such that the sum of twice the first and three times the second is 7 more than twice the third.

Applications

Solve.

9. An art show charges $12.50 for admission and sells 4-inch by 6-inch postcards of works by the featured artists for $2.75 each. Brooke attends the art show and spends a total of $37.25 on admission and postcards. The situation can be modeled by $37.25 = $12.50 + $2.75p.

   a. The unknown value is represented by the variable p in the equation. What is the unknown value in this situation?

   b. Solve the equation for the variable.

   c. What does the answer to part b. mean? Write a complete sentence.

10. Robin is in charge of purchasing desserts for a dinner party that her nonprofit organization is throwing. She decides to buy a cake and several specialty cupcakes from Barbara's Bombtastic Bakery. She needs to buy one 8-inch round cake which costs $19.50. She has $45 to spend and will spend the leftover amount on cupcakes, which are $8.50 for a box of 4. How many boxes of cupcakes can Robin purchase?

    a. What is the unknown value in this problem? Let the variable c represent this unknown value.

    b. Write an equation to represent this situation.

    c. Solve the equation for the variable.

    d. What does the answer to part c. mean? Write a complete sentence.

Writing & Thinking

11.   a. How would you represent four consecutive odd integers?

    b. How would you represent four consecutive even integers?

    c. Are these representations the same? Explain.

9.6 Applications: Distance-Rate-Time, Interest, Average, and Cost

▶ Watch and Work

Watch the video for Example 6 in the software and follow along in the space provided.

Example 6 Application: Calculating Cost

A jeweler paid $350 for a ring. He wants to price the ring for sale so that he can give a 30% discount on the marked selling price and still make a profit of 20% on his cost. What should be the marked selling price of the ring?

Solution

Now You Try It!

Use the space provided to work out the solution to the next example.

Example A Application: Calculating Cost

A used car salesman paid $1500 for a car. He plans to sell the car at a 25% discount on the marked selling price, but still wants to make a profit of 15% on his cost. What should be the marked selling price of the car?

9.6 Exercises

Concept Check

True/False. Determine whether each statement is true or false. If a statement is false, explain how it can be changed so the statement will be true. (**Note:** There may be more than one acceptable change.)

1. When using the formula $I = Pr$, the value of r should be written as a percent.

2. In the distance-rate-time formula $d = r \cdot t$, the value t stands for the time spent traveling.

3. Profit can be determined by subtracting the cost from the selling price.

4. The concept of average can be used to find unknown numbers.

Applications

Solve.

5. Jamie plans to take the scenic route from Los Angeles to San Francisco. Her GPS tells her it is a 420-mile trip. If she figures her average speed will be 48 mph, how long will the trip take her?

6. Amanda invests $25,000, part at 5% and the rest at 6%. The annual return on the 5% investment exceeds the annual return on the 6% investment by $40. How much did she invest at each rate?

7. A particular style of shoe costs the dealer $81 per pair. At what price should the dealer mark them so he can sell them at a 10% discount off the selling price and still make a 25% profit?

8. Marissa has five exam scores of 75, 82, 90, 85, and 77 in her chemistry class. What score does she need on the final exam to have an average grade of 80 (and thus earn a grade of B)? (All exams have a maximum of 100 points.)

Writing & Thinking

Each of the following problems is given with an incorrect answer. Explain how you can tell that the answer is incorrect without needing to solve the problem or do any algebra; then, solve the problem correctly.

9. The perimeter of an isosceles triangle is 16 cm. Since the triangle is isosceles, two sides have the same length; the third side is 2 cm shorter than one of the two equal sides. Find the length of one of the two equal sides.
Incorrect answer: 9 cm

10. Leela found a used textbook, which was marked down 50% from the price of the new textbook. If the used textbook cost $60, how much did the new textbook cost? **Incorrect answer: $90**

9.7 Solving Linear Inequalities in One Variable

1. The set of all real numbers between a and b is called an _____.

Type of Interval	Algebraic Notation	Interval Notation	Graph
Open Interval	_____	(a, b)	
Closed Interval	$a \leq x \leq b$	_____	
_____	$\begin{cases} a \leq x < b \\ a < x \leq b \end{cases}$	$[a, b)$ $(a, b]$	
Open Interval	$\begin{cases} x > a \\ x < b \end{cases}$	_____	
Half-open Interval	_____	$[a, \infty)$ $(-\infty, b]$	

Table 1

2. In an **open interval**, _____.

3. In a **closed interval**, _____.

4. In a **half-open interval**, _____.

5. **Linear inequalities** are inequalities that _____.

6. A **solution** to an inequality is any number that _____.

9.7 Solving Linear Inequalities in One Variable

Addition Principle for Solving Linear Inequalities

If A and B are algebraic expressions and C is a real number, then _____

_____

and

_____

_____

(If a real number is added to both sides of an inequality, the new inequality is _____

_____.)

PROPERTIES

Multiplication Principle for Solving Linear Inequalities

If A and B are algebraic expressions and C is a positive real number, then _____

_____

and

_____

_____

If A and B are algebraic expressions and C is a negative real number, then _____

_____

and

_____

_____

(In other words, if both sides are multiplied by a _____

_____.)

PROPERTIES

Solving Linear Inequalities

1. Combine _____

2. Use the addition principle of inequality to _____

_____

3. Use the multiplication (or division) principle of inequality to multiply (or divide) both sides by the coefficient of the variable so that _____
   If this coefficient is negative, _____

4. A quick (and generally satisfactory) check is to _____

_____
   If the statement is false, _____

PROCEDURE

▶ Watch and Work

Watch the video for Example 9 in the software and follow along in the space provided.

Example 9 Solving Linear Inequalities

Solve the inequality $6x + 5 \leq -1$ and graph the solution set. Write the solution set using interval notation.

Solution

Now You Try It!

Use the space provided to work out the solution to the next example.

Example A Solving Linear Inequalities

Solve the inequality $4x+8>-16$ and graph the solution set. Write the solution set using interval notation.

9.7 Exercises

Concept Check

True/False. Determine whether each statement is true or false. If a statement is false, explain how it can be changed so the statement will be true. (**Note:** There may be more than one acceptable change.)

1. If only one endpoint is included in an interval, it is called a half-open interval.

2. When both sides of a linear inequality are multiplied by a negative constant, the sense of the inequality should stay the same.

3. To check the solution set of a linear inequality, every solution in the solution set must be checked in the original inequality.

4. The infinity symbol ∞ does not represent a specific number.

Practice

Graph each interval on a real number line and tell what type of interval it is.

5. $x \leq -3$

6. $-1.5 \leq x < 3.2$

Solve each inequality and graph the solution set. Write each solution set using interval notation.

7. $x + 1 > 5$

8. $-2x \geq 6$

9. $4x - 7 \geq 9$

10. $5x + 6 \geq 2x - 2$

Applications

Solve.

11. A statistics student has grades of 82, 95, 93, and 78 on four hour-long exams. He must average 90 or higher to receive an A for the course. What scores can he receive on the final exam and earn an A if:

   a. The final is equivalent to a single hour-long exam (100 points maximum)?

   b. The final is equivalent to two hourly exams (200 points maximum)?

12. Allison is ordering boxes of 24 tea bags from a website. The website is having a promotion where each box of tea comes with 2 free sample packs, and each sample pack contains 3 tea bags. If Allison has an empty container that holds 150 tea bags, what is the largest number of boxes of tea Allison can order and not overfill the container?

Writing & Thinking

13. a. Write a list of three situations where inequalities might be used in daily life.

    b. Illustrate theses situations with algebraic inequalities and appropriate numbers.

9.8 Compound Inequalities

> **Union and Intersection**
>
> The **union** (symbolized $\cup$, as in $A \cup B$) of two (or more) sets is _____
> _____
>
> The **intersection** (symbolized $\cap$, as in $A \cap B$) of two (or more) sets is _____
> _____
>
> The word **or** is used to indicate _____ and the word **and** is used to indicate _____.
>
> **DEFINITION**

1. If the elements in a set can be counted, the set is said to be _____. If the elements cannot be counted, the set is said to be _____.

▶ Watch and Work

Watch the video for Example 10 in the software and follow along in the space provided.

Example 10   Solving Compound Inequalities Containing OR

Solve the compound inequality $3x + 4 \geq 1$ or $2x - 7 \leq -3$ and graph the solution set. Write the solution set using interval notation.

Solution

✏️ Now You Try It!

Use the space provided to work out the solution to the next example.

Example A Solving Compound Inequalities Containing OR

Solve the inequality $4x - 3 \geq 1$ or $5x + 2 < 12$ and graph the solution set. Write the solution set using interval notation.

9.8 Exercises

Concept Check

True/False. Determine whether each statement is true or false. If a statement is false, explain how it can be changed so the statement will be true. (**Note:** There may be more than one acceptable change.)

1. The union of two sets is the set of all elements that belong to both sets.

2. The intersection of two sets is the set of elements that belong to just one set or the other, but not both.

3. A null set contains no elements.

4. The solution set of a compound inequality containing and is the union of the solution sets of the two inequalities.

Practice

5. Find the union and intersection of $A = \{2,4,6,8\}$ and $B = \{1,2,3,4\}$.

6. Graph the set $\{x \mid x > 3 \text{ or } x \leq 2\}$ on a real number line.

9.8 Exercises

Solve each compound inequality and graph its solution set. Write each solution set using interval notation.

7. $\{x \mid x+3 > 2 \text{ and } x-1 < 5\}$

8. $\{x \mid x < -1 \text{ or } 2x+1 \geq 3\}$

9. $-4 < x+5 < 6$

Name: Date: **311**

9.9 Absolute Value Equations

> **Absolute Value**
> The **absolute value** of a number is _____
>
> DEFINITION

> **Solving Absolute Value Equations**
> For $c > 0$:
> a. If $|x| = c$, then _____
> b. If $|ax + b| = c$, then _____
>
> DEFINITION

> **Solving Equations with Two Absolute Value Expressions**
> If $|a| = |b|$, then _____
>
> More generally,
>
> if $|ax + b| = |cx + d|$, then _____
>
> DEFINITION

▶ Watch and Work

Watch the video for Example 2 in the software and follow along in the space provided.

Example 2 Solving Equations with Two Absolute Value Expressions

Solve: $|x + 5| = |2x + 1|$

Solution

✏️ Now You Try It!

Use the space provided to work out the solution to the next example.

Example A Solving Equations with Two Absolute Value Expressions

Solve: $|4x-5| = |3x-16|$

9.9 Exercises

Concept Check

True/False. Determine whether each statement is true or false. If a statement is false, explain how it can be changed so the statement will be true. (**Note:** There may be more than one acceptable change.)

1. Equations involving absolute value can only have one solution.

2. If two numbers have the same absolute value, they must be equal to each other.

3. There is no number that has a negative absolute value.

4. If $|a| = |b|$, we can only rewrite it as $a = b$.

Practice

Solve each absolute value equation.

5. $|z| = -\dfrac{1}{5}$

6. $|y+5| = -7$

7. $|-2x+1| = -3$

8. $3\left|\dfrac{x}{3}+1\right| - 5 = -2$

9. $\left|\dfrac{x}{3}-4\right| = \left|\dfrac{5x}{6}+1\right|$

9.9 Exercises

9.10 Absolute Value Inequalities

Algebraic Notation	Graph	Interval Notation
$\|x\| < 3$ _____ (the intersection)	3 units, 3 units; number line from −3 to 3	_____

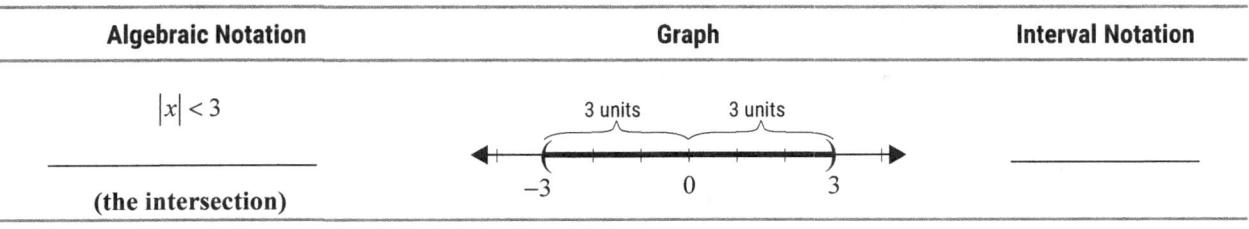

Table 1

Solving Absolute Value Inequalities with < (or ≤)

For $c > 0$:

a. If $|x| < c$, then _____

b. If $|ax + b| < c$, then _____

The inequalities in **a.** and **b.** are also true if < is _____

DEFINITION

Algebraic Notation	Graph	Interval Notation
$\|x\| > 3$ _____ (the union)	3 units, 3 units; number line from −3 to 3	_____

Table 2

Solving Absolute Value Inequalities with > (or ≥)

For $c > 0$:

a. If $|x| > c$, then _____

b. If $|ax + b| > c$, then _____

The inequalities in **a.** and **b.** are true if > is _____

DEFINITION

9.10 Absolute Value Inequalities

▶ Watch and Work

Watch the video for Example 9 in the software and follow along in the space provided.

Example 9 Solving Absolute Value Inequalities

Solve the absolute value inequality and graph the solution set: $|2x-5|-5 \geq 4$

Solution

✏ Now You Try It!

Use the space provided to work out the solution to the next example.

Example A Solving Absolute Value Inequalities

Solve the absolute value inequality $|3x+2|+1 \geq 8$ and graph the solution set. Write the solution set using interval notation.

9.10 Exercises

Concept Check

True/False. Determine whether each statement is true or false. If a statement is false, explain how it can be changed so the statement will be true. (**Note:** There may be more than one acceptable change.)

1. If the solution is a union, there are two statements or inequalities, both of which must be true.

2. If the solution to a compound inequality is $-4 < x < 6$, then the solution is a union.

3. For a number to have absolute value greater than 2, its distance from 0 must be less than 2.

4. The inequality $|2x + 9| < -2$ has no solution.

Practice

Solve each of the absolute value inequalities and graph the solution sets. Write each solution using interval notation.

5. $|x| \geq -2$

6. $|x - 3| > 2$

7. $|x + 2| \leq -4$

8. $|3x + 4| - 1 < 0$

9. $4 \leq |3x + 1| - 6$

318 9.10 Exercises

10. $3|4x+5|-5>10$

Writing & Thinking

A set of real numbers is described. **a**. Sketch a graph of the set on a real number line. **b**. Represent each set using absolute value notation. **c**. Represent each set using interval notation. If the set is one interval, state what type of interval it is.

11. The set of real numbers between −10 and 10, inclusive

12. The set of real numbers within 7 units of 4

Chapter 9 Project

A Linear Vacation

An activity to demonstrate the importance of solving linear equations in real life.

The process of finding ways to use math to solve real-life scenarios is called mathematical modeling. In the following activity you will be using linear equations to model some real-life scenarios that arise during a family vacation.

For each question, write a linear equation in one variable and then solve.

1. Penny and her family went on vacation to Florida and decided to rent a car to do some sightseeing. The cost of the rental car was a fixed price per day plus $0.29 per mile. When she returned the car, the bill was $209.80 for three days and they had driven 320 miles. What was the fixed price per day to rent the car?

2. Penny's son Chase wanted to go to the driving range to hit some golf balls. Penny gave the pro-shop clerk $60 for three buckets of golf balls and received $7.50 in change. What was the cost of each bucket?

3. Penny's family decided to go to the Splash Park. They purchased two adult tickets and two child tickets. The adult tickets were $1\frac{1}{2}$ times the price of the child tickets and the total cost for all four tickets was $130. What was the cost of each type of ticket?

4. Penny's family went shopping at a nearby souvenir shop where they decided to buy matching T-shirts. If they bought four T-shirts and a $9.99 bottle of sunscreen for a total cost of $89.51, before tax, how much did each T-shirt cost?

5. Penny and her family went out to eat at a local restaurant. Three of them ordered a shrimp basket, but her daughter Meghan ordered a basket of chicken tenders, which was $4.95 less than the shrimp basket. If the total order before tax was $46.85, what was the price of a shrimp basket?

6. While on the beach, Penny and her family decided to play a game of volleyball. Penny and her son beat her husband and daughter by two points. If the combined score of both teams was 40, what was the score of the winning team?

Chapter 9 Project

Breaking Even

An activity to demonstrate the use of linear equations and inequalities in business.

In manufacturing, the production cost for an item usually has two components: a fixed cost and a variable cost. You can think of the fixed cost as money that must be spent to operate the business regardless of production level. Examples of fixed cost include paying the rent or mortgage for the manufacturing facility or insurance on the property. The variable cost reflects the funds that must be spent to produce one unit of the product. Variable cost should account for things like raw materials and labor costs.

1. A guitar manufacturer has daily fixed cost of $12,000 and each guitar costs $230 to build.
   a. Determine the total cost of producing 10 guitars in one day.
   b. How many guitars were produced in a day when the total cost was $21,890?
   c. What is the minimum number of guitars produced in a day that will make the total cost exceed $50,000?

2. The revenue for a manufacturer is the income generated from selling products. Revenue is defined as the price per unit times the number of units sold. The guitar manufacturer from our previous problem can sell each guitar for $400.
   a. Determine the revenue when 10 guitars are sold.
   b. What is the minimum number of guitars that the manufacturer must sell in a day so that revenue is at least $50,000?

3. Profit is defined as revenue minus cost. The break-even point is the production level at which the profit is exactly zero. Above that level, the manufacturer returns a profit. Determine the break-even point for this guitar manufacturer. In other words, find the number of guitars that must be manufactured and sold in a day to cover all the manufacturing cost.

4. If the manufacturer could decrease fixed cost from $12,000 to $10,000, would you expect the break-even point to go up or down? Explain. Use a linear equation to verify your prediction.

5. Assume that the manufacturer keeps fixed cost at $12,000 with a production cost of $230 per guitar. What should the sale price be to keep a break-even point of 80 units?

6. Write a linear inequality that represents the company returning a positive profit if the fixed cost is $12,000 with a production cost of $230 per guitar and a sale price of $400 per guitar.

7. Solve the inequality found in Problem 6. Round to the nearest integer and write the solution set in interval notation. Explain what this solution set represents.

8. Based on the solution set found in Problem 7, is there a limit to the number of guitars that can be manufactured and sold and still return a profit? Do you think this is realistic? Explain your answer.

CHAPTER 10

Graphing Linear Equations and Inequalities

10.1 The Cartesian Coordinate System

10.2 Graphing Linear Equations in Two Variables

10.3 Slope-Intercept Form

10.4 Point-Slope Form

10.5 Introduction to Functions and Function Notation

10.6 Graphing Linear Inequalities in Two Variables

CHAPTER 10 PROJECTS
What's Your Car Worth?

Demand and It Shall Be Supplied

Connections

Rideshare companies such as Uber and Lyft are becoming increasingly more popular in the United States. The website carsurance.net reported that one in every four people in America used a rideshare service at least once per month in 2020. With these numbers, the industry is projected to surpass $220 billion dollars in value by the year 2025.

Prices for the rideshare services vary slightly from company to company, and different markets have their own pricing formulas. However, the general idea is the same: a customer pays a fixed fee to request a ride and then each mile driven costs a certain amount.

Suppose that in Columbus, Ohio, the approximate pricing formulas are as follows: Uber charges a flat fee of $2.80 plus $1.60 per mile and Lyft charges a flat fee of $3.80 plus $1.30 per mile. For very short rides, Uber seems to cost less than Lyft. But as the ride gets longer, Lyft becomes a better deal. The cost of each service as a function of miles is graphed here.

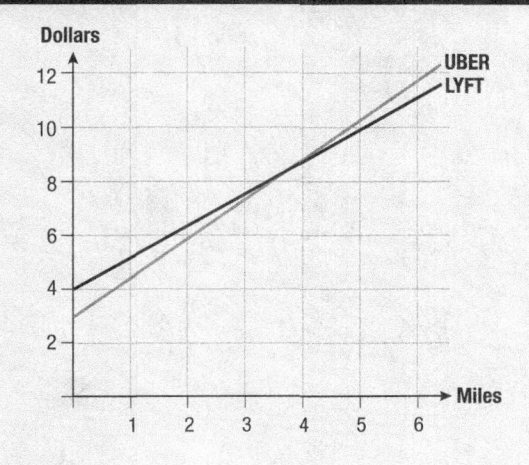

Given this information, could you determine what length a trip must be for a ride with Lyft to be less expensive than a ride with Uber?

10.1 The Cartesian Coordinate System

1. Descartes based his system on a relationship between _____ in a plane and _____ of real numbers.

2. In the ordered pair (x, y), x is called the _____ and y is called the _____.

3. In an ordered pair of the form (x, y), the _____ is called the **independent variable** and the _____ is called the **dependent variable**.

4. The Cartesian coordinate system relates algebraic equations and ordered pairs to geometry. In this system, two number lines intersect at right angles and separate the plane into four _____. The **origin**, designated by the ordered pair $(0, 0)$, is _____. The horizontal number line is called the _____ or _____. The vertical number line is called the _____ or _____.

One-to-One Correspondence

_____

_____

DEFINITION

▶ Watch and Work

Watch the video for Example 4 in the software and follow along in the space provided.

Example 4 Finding Ordered Pairs

Complete the table so that each ordered pair will satisfy the equation $y = -3x + 1$.

x	y	(x, y)
0		
	4	
$\frac{1}{3}$		
3		

Solution

✏ Now You Try It!

Use the space provided to work out the solution to the next example.

Example A Finding Ordered Pairs

Complete the table so that each ordered pair will satisfy the equation $y = -3x + 2$.

x	y	(x, y)
0		
	1	
−2		
	0	

10.1 Exercises

Concept Check

True/False. Determine whether each statement is true or false. If a statement is false, explain how it can be changed so the statement will be true. (**Note:** There may be more than one acceptable change.)

1. The graph of every ordered pair that has a positive x-coordinate and a negative y-coordinate can be found in Quadrant IV.

2. To find the y-value that corresponds with $x = 2$, substitute 2 for x into the given equation and solve for y.

3. If $(-7, 3)$ is a solution of $y = 3x + 24$, then $(-7, 3)$ satisfies $y = 3x + 24$.

4. If point $A = (0, 4)$, then point A lies on the x-axis.

Practice

List the set of ordered pairs corresponding to the points on the graph.

5.

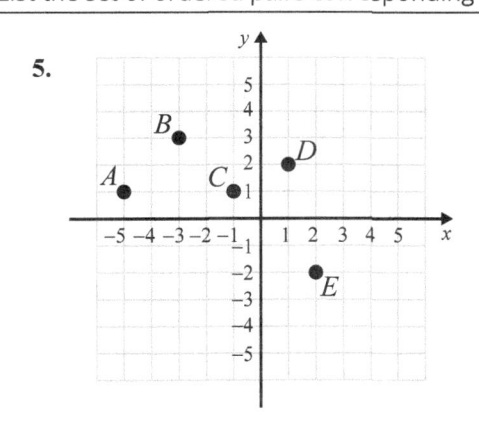

10.1 Exercises

Plot each set of ordered pairs and label the points.

6. $\{A(4, -1), B(3, 2), C(0, 5), D(1, -1), E(1, 4)\}$

Determine the missing coordinate in each of the ordered pairs so that the point will satisfy the equation given.

7. $x - 2y = 2$

   a. $(0, \_\_)$

   b. $(4, \_\_)$

   c. $(\_\_, 0)$

   d. $(\_\_, 3)$

Complete the table so that each ordered pair will satisfy the given equation. Plot the resulting sets of ordered pairs.

8. $y = 2x - 3$

x	y
0	
	-1
-2	
	3

Determine which, if any, of the ordered pairs satisfy the given equation.

9. $2x - 3y = 7$
   a. $(1, 3)$
   b. $\left(\dfrac{1}{2}, -2\right)$
   c. $\left(\dfrac{7}{2}, 0\right)$
   d. $(2, 1)$

The graph of a line is shown. List any three points on the line. (There is more than one correct answer.)

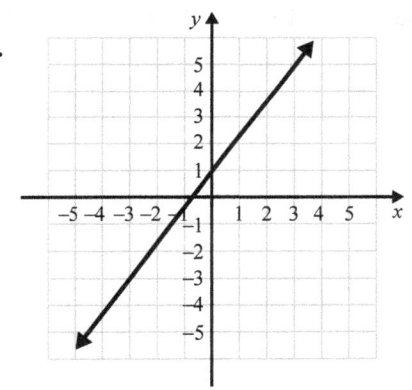

10.

Applications

Solve.

11. At one point in 2017, the exchange rate from US dollars to Euros was $E = 0.85D$ where E is Euros and D is dollars.

    a. Make a table of ordered pairs for the values of D and E if D has the values $100, $200, $300, $400, and $500.

    b. Plot the points corresponding to the ordered pairs.

12. Consider the equation $F = \frac{9}{5}C + 32$, where C is temperature in degrees Celsius and F is the corresponding temperature in degrees Fahrenheit.

    a. Make a table of ordered pairs for the values of C and F if C has the values $-20°$, $-10°$, $-5°$, $0°$, $5°$, $10°$, and $15°$.

    b. Plot the points corresponding to the ordered pairs.

10.2 Graphing Linear Equations in Two Variables

Standard Form of a Linear Equation

Any equation of the form

_____

where A, B, and C are real numbers and A and B are not both equal to 0, is called the standard form of a linear equation.

DEFINITION

Graphing a Linear Equation in Two Variables

1. Locate any two points that _____
2. _____
3. _____
4. To check: Locate a third point that _____

PROCEDURE

▶ Watch and Work

Watch the video for Example 2 in the software and follow along in the space provided.

Example 2 Graphing a Linear Equation in Two Variables

Graph: $2x + 3y = 6$

Solution

✏️ Now You Try It!

Use the space provided to work out the solution to the next example.

Example A Graphing a Linear Equation in Two Variables

Graph: $3x + 2y = 6$

Intercepts

1. To find the y-intercept (where the line crosses the y-axis),
   _____

2. To find the x-intercept (where the line crosses the x-axis),
   _____

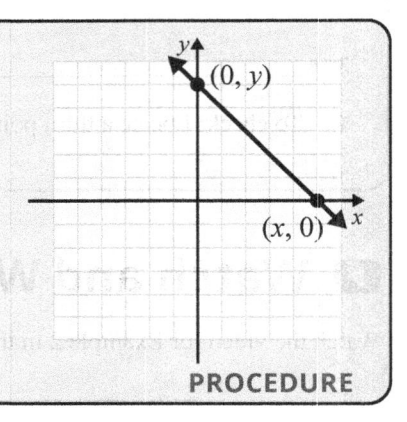

PROCEDURE

Horizontal and Vertical Lines

For real numbers a and b, the graph of _____

DEFINITION

10.2 Exercises

Concept Check

True/False. Determine whether each statement is true or false. If a statement is false, explain how it can be changed so the statement will be true. (**Note:** There may be more than one acceptable change.)

1. The y-intercept is the point where a line crosses the y-axis.

2. The terms ordered pair and point are used interchangeably.

3. A horizontal line does not have a y-intercept.

4. All x-intercepts correspond to an ordered pair of the form $(0, y)$.

Practice

Graph each linear equation by locating at least two ordered pairs that satisfy the given equation.

5. $x + y = 3$

7. $y = -3$

6. $x = 1$

Graph each linear equation by locating the x-intercept and the y-intercept.

8. $y = 4x - 10$

9. $3x - 7y = -21$

Applications

Solve.

10. The amount of potassium in a clear bottle of a popular sports drink declines over time when exposed to the UV lights found in most grocery stores. The amount of potassium in a container of this sports drink is given by the equation $y = -30x + 360$, where y represents the mg of potassium remaining after x days on the shelf. Find both the x-intercept and y-intercept, and interpret the meaning of each in the context of this problem.

11. Mr. Adler has found that the grade each student gets in his Introductory Algebra course directly correlates with the amount of time spent doing homework, and is represented by the equation $y = 7x + 30$, where y represents the numerical score the student receives on an exam (out of 100 points) after spending x hours per week doing homework. Find the y-intercept and interpret its meaning in this context.

Writing & Thinking

12. Explain, in your own words, why it is sufficient to find the x-intercept and y-intercept to graph a line (assuming that they are not the same point).

13. Explain, in your own words, how you can determine if an ordered pair is a solution to an equation.

10.3 Slope-Intercept Form

1. For a line, the _____ is called the **slope of the line**.

> **Slope**
>
> Let $P_1(x_1, y_1)$ and $P_2(x_2, y_2)$ be two points on a line. The slope can be calculated as follows.
>
> $$\text{slope} = \underline{\hspace{3cm}}$$
>
> **Note:** _____ is standard notation for representing the slope of a line.
>
> FORMULA

▶ Watch and Work

Watch the video for Example 2 in the software and follow along in the space provided.

Example 2 Finding the Slope of a Line

Find the slope of the line that contains the points $(1, 3)$ and $(5, 1)$, and then graph the line.

Solution

10.3 Slope-Intercept Form

✏ Now You Try It!

Use the space provided to work out the solution to the next example.

Example A Finding the Slope of a Line

Find the slope of the line that contains the points $(0, 5)$ and $(4, 2)$, and then graph the line.

Positive and Negative Slope

Lines with positive slope go _____

Lines with negative slope go _____

DEFINITION

Horizontal and Vertical Lines

The following two general statements are true for horizontal and vertical lines.

1. For horizontal lines (of the form _____

2. For vertical lines (of the form _____

DEFINITION

Slope-Intercept Form

_____ is called the slope-intercept form for the equation of a line, where m is the slope and $(0, b)$ is the y-intercept.

DEFINITION

10.3 Exercises

Concept Check

True/False. Determine whether each statement is true or false. If a statement is false, explain how it can be changed so the statement will be true. (**Note:** There may be more than one acceptable change.)

1. If the y-intercept and the slope of a line are given, there is enough information to write the equation of the line.

2. When using the slope formula, the slope of a line changes if the order of the points is reversed.

3. A line that falls (decreases) from left to right has a negative slope.

4. The line that represents the equation $y = 2x + 4$ has a y-intercept of $(0, 4)$.

Practice

Find the slope of the line determined by each pair of points.

5. $(1, -2); (1, 4)$

6. $(-3, 7); (4, -1)$

Determine whether the equation $x = -3$ represents a horizontal line or a vertical line and give its slope.

7. $x = -3$

338 10.3 Exercises

Write each equation in slope-intercept form. Find the slope and y-intercept, and then use them to draw the graph.

8. $y = 2x - 1$

9. $3y - 9 = 0$

Find an equation in slope-intercept form for the line passing through (0,3) with the slope $m = -\dfrac{1}{2}$.

10. $(0, 3); m = -\dfrac{1}{2}$

Applications

Solve.

11. John bought his new car for $35,000 in the year 2022. He knows that the value of his car has depreciated linearly. If the value of the car in 2025 was $23,000, what was the annual rate of depreciation of his car? Show this information on a graph. (When graphing, use years as the x-coordinates and the corresponding values of the car as the y-coordinates.)

12. The number of people in the United States with cell phones was about 198 million in 2011 and about 232 million in 2016. If the growth in the usage of cell phones was linear, what was the approximate rate of growth per year from 2011 to 2016. Show this information on a graph. (When graphing, use years as the x-coordinates and the corresponding numbers of users as the y-coordinates.)[1]

---

1 Source: www.statista.com/statistics/231612/number-of-cell-phone-users-usa

Writing & Thinking

13. a. Explain in your own words why the slope of a horizontal line must be 0.

    b. Explain in your own words why the slope of a vertical line must be undefined.

Name: _____   Date: _____   **341**

10.4 Point-Slope Form

Point-Slope Form

An equation of the form

_____

is called the point-slope form for the equation of a line that _____
_____

DEFINITION

Finding the Equation of a Line Given Two Points

To find the equation of a line given two points on the line:

1. Use the formula
   _____

2. Use this slope, m, and _____
   _____

PROCEDURE

▶ Watch and Work

Watch the video for Example 4 in the software and follow along in the space provided.

Example 4 Finding Equations of Lines Using a Graph

Write an equation in standard form for the following line.

Solution

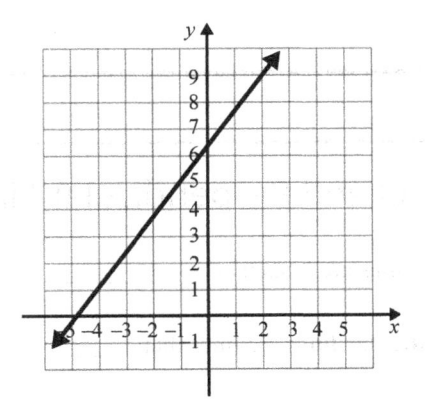

10.4 Point-Slope Form

✏️ Now You Try It!

Use the space provided to work out the solution to the next example.

Example A Finding Equations of Lines Using a Graph

Write an equation in standard form for the following line.

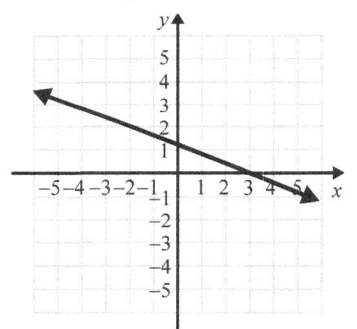

Parallel and Perpendicular Lines

Parallel lines are lines that _____ and who have the _____.

Perpendicular lines are lines that _____ and whose slopes are _____. Horizontal lines are perpendicular to _____.

DEFINITION

10.4 Exercises

Concept Check

True/False. Determine whether each statement is true or false. If a statement is false, explain how it can be changed so the statement will be true. (**Note:** There may be more than one acceptable change.)

1. Given two perpendicular lines (neither of which have slope 0), we know that one has a positive slope and the other has a negative slope.

2. If Line 2 is parallel to Line 3, then the slope of Line 2 equals the slope of Line 3.

3. A line perpendicular to a horizontal line has a slope that is undefined.

4. All pairs of lines are either parallel or perpendicular.

Practice

Find **a.** the slope, **b.** a point on the line, and **c.** the graph of the line for the following equations in point-slope form.

5. $y - 1 = 2(x - 3)$

Find an equation in standard form for the line passing through the given point with the given slope. Graph the line.

6. $(3, 4)$; $m = 3$

Find an equation in slope-intercept form for the line passing through the two given points.

7. $(-5, 1)$; $(2, 0)$

10.4 Exercises

Find an equation in standard form for the line shown.

8.

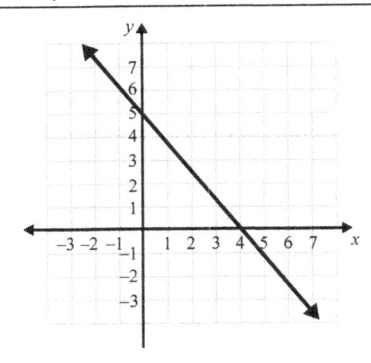

Find an equation in slope-intercept form that satisfies each set of conditions.

9. Find an equation in slope-intercept form for the horizontal line through the point $(-2, 6)$.

Determine whether the pair of lines is **a.** parallel, **b.** perpendicular, or **c.** neither. Graph both lines. (**Hint:** Compare the slopes.)

10. $\begin{cases} y = -2x + 3 \\ y = -2x - 1 \end{cases}$

Applications

Solve.

11. The cost for an airline to fly from Raleigh, NC, to Nashville, TN, is $5000. The airline charges $100 for the one-way ticket from Raleigh to Nashville.

   a. Find an equation for the profit P made by the airline on this one-way flight if they sell t tickets.

   b. Use the equation found in part a. to determine the number of tickets that must be sold for the airline to "break even;" that is, for the profit to be equal to 0?

12. Natalie invested some money in a simple interest savings fund. After 2 years, she earned $120 in interest. After 5 years, she earned $300 in interest.

   a. Write two ordered pairs from the information given where x represents the time in years and y represents the amount of interest earned.

   b. Find the slope of the line which contains the two ordered pairs from part a.

   c. Write the point-slope equation that models the situation.

   d. Rewrite this equation in $y = mx + b$ form.

10.5 Introduction to Functions and Function Notation

Relation, Domain, and Range

A **relation** is a set of _____

The **domain**, *D*, of a relation is the set of _____

The **range**, *R*, of a relation is the set of _____

DEFINITION

Functions

A **function** is a relation in which _____
_____

DEFINITION

Vertical Line Test

If any vertical line intersects the graph of a relation at more than one point, then the relation is _____

PROCEDURE

Linear Function

A linear function is a function represented by an equation of the form

_____

The domain of a linear function is _____

_____

DEFINITION

1. In function notation, instead of writing *y*, write _____ , which is read "*f* of *x*."

▶ Watch and Work

Watch the video for Example 6 in the software and follow along in the space provided.

Example 6 Evaluating Functions

For the function $g(x) = 4x + 5$, find:

a. $g(2)$
b. $g(-1)$
c. $g(0)$

Solution

✏ Now You Try It!

Use the space provided to work out the solution to the next example.

Example A Evaluating Functions

For the function $g(x) = 3x - 2$, find:

a. $g(3)$
b. $g(-2)$
c. $g(0)$

10.5 Exercises

Concept Check

True/False. Determine whether each statement is true or false. If a statement is false, explain how it can be changed so the statement will be true. (**Note:** There may be more than one acceptable change.)

1. If the domain of a linear function is not explicitly stated, the implied domain is the set of all values of x that produce real values for y.

2. A relation is a function in which each domain element has exactly one corresponding range element.

3. In a function, the range elements can have more than one corresponding domain element.

4. If $s = \{(1, -6), (3, 5), (4, 0), (1, 2)\}$, then s is a function.

Practice

List the sets of ordered pairs that correspond to the points. State the domain and range and indicate if the relation is a function.

5.

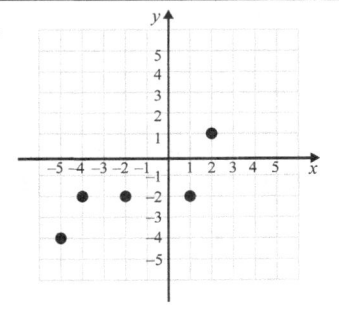

Graph the relation. State the domain and range and indicate which of the relation is a function.

6. $h = \{(1, -5), (2, -3), (-1, -3), (0, 2), (4, 3)\}$

10.5 Exercises

Use the vertical line test to determine whether the graph represents a function. State the domain and range using interval notation.

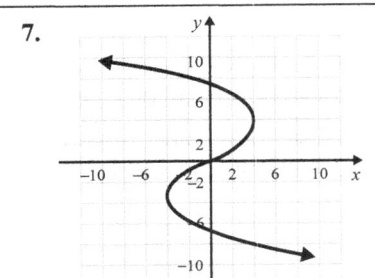

7.

State the domain of the function.

8. $h(x) = \dfrac{7}{3x}$

Find the values of the function as indicated.

9. $F(x) = 6x^2 - 10$

   a. $F(0)$
   b. $F(-4)$
   c. $F(4)$

Applications

Solve.

10. A nurse hangs a 1000-milliliter IV bag which is set to drip at 120 milliliters per hour. Create a model of this situation to represent the amount of IV solution left in the bag after x hours.

    a. The y-intercept is the amount of IV solution in the bag initially (time = 0). What is the y-intercept?
    b. The slope is equal to the rate that the IV solution is dispensed per hour. What is the slope? (**Hint:** Consider whether the amount of IV solution in the bag is increasing or decreasing and how this would affect the slope.)
    c. Write an equation in slope-intercept form to model this situation.
    d. Write the equation from part c. using function notation.
    e. State the domain and range of the function.
    f. State any additional restrictions that should be made on the domain for it to make sense in the context of this problem.
    g. How much IV solution is left in the bag after 5 hours?

10.6 Graphing Linear Inequalities in Two Variables

Half-plane   A straight line separates _____

The points on one side of the line are in _____

_____

Boundary line   The line _____

Closed half-plane   If the boundary line is _____

_____

Open half-plane   If the boundary line is _____

_____

Graphing Linear Inequalities

1. First, graph _____ (_____ if the inequality is _____ if the inequality is _____).

2. Next, determine which _____

   Method 1
   a. Test any one _____
   b. If the test-point satisfies the inequality, shade the _____
   _____

   Note: The point $(0, 0)$, if it is not on the boundary line, is usually the easiest point to test.

   Method 2
   a. Solve the inequality for y _____
   b. If the solution shows y _____
   c. If the solution shows y _____

   Note: If the boundary line is vertical, then _____

   If the solution shows _____

   If the solution shows _____

3. The shaded half-plane (and the line if it is solid) _____

PROCEDURE

▶ Watch and Work

Watch the video for Example 2 in the software and follow along in the space provided.

Example 2 Graphing Linear Inequalities

Graph the solution set to the inequality $y > 2x$.

Solution

✏ Now You Try It!

Use the space provided to work out the solution to the next example.

Example A Graphing Linear Inequalities

Graph the solution set to the inequality $y < 3x$.

10.6 Exercises

Concept Check

True/False. Determine whether each statement is true or false. If a statement is false, explain how it can be changed so the statement will be true. (**Note:** There may be more than one acceptable change.)

1. A solid boundary line indicates that the points on that line are included in the solution.

2. If the solution set is an open half-plane, then the boundary line is included in the solution.

3. The boundary line is solid when the inequality uses a < or > symbol.

4. The slope of an inequality is used to determine whether the boundary line is included in the solution.

Practice

Graph the solution set of each of the linear inequalities.

5. $x + y \leq 7$

6. $5x - y < 4$

7. $x + 4 \geq 0$

8. $\dfrac{1}{2}x - y > 1$

9. $2x - \dfrac{4}{3}y > 8$

Applications

Solve.

10. The grade for a 1-credit-hour survey class is based on an exam and a project, which are worth a maximum of 50 points each. The sum of the two scores must be at least 75 points for a student to earn a passing grade.

   a. Let the amount of points earned on the exam be represented by the variable x and the amount of points earned on the project be represented by the variable y. Create a linear inequality to describe the solution set for a passing grade.

   b. Graph the linear inequality from part a.

   c. A student earns 45 points on their final exam and 22 points on their project. Plot this point on the graph. Did this student earn a passing grade?

   d. Are there any points in the solution set which do not make sense for this situation?

Writing & Thinking

11. Explain in your own words how to test to determine which side of the graph of an inequality should be shaded.

12. Describe the difference between a closed and an open half-plane.

Chapter 10 Project

What's Your Car Worth?

An activity to demonstrate the use of linear models in real life.

When buying a new car, there are a number of things to keep in mind: your monthly budget, length of the warranty, routine maintenance costs, potential repair costs, cost of insurance, etc.

One thing you may not have considered is the *depreciation*, or reduction in value, of the car over time. If you like to purchase a new car every 3 to 5 years, then the *retention value* of a car, or the portion of the original price remaining, is an important factor to keep in mind. If your new car depreciates in value quickly, you may have to settle for less money if you choose to resell it later or trade it in for a new one.

Below is a table of original Manufacturer's Suggested Retail Price (MSRP) values and the anticipated retention value after 3 years for three 2022 mid-price car models.

Car Model	2022 MSRP	Expected Value in 2025	Rate of Depreciation (slope)	Linear Equation
Mini Cooper	$28,600	$16,390		
Toyota Camry	$25,845	$13,101		
Ford Taurus	$30,230	$13,244		

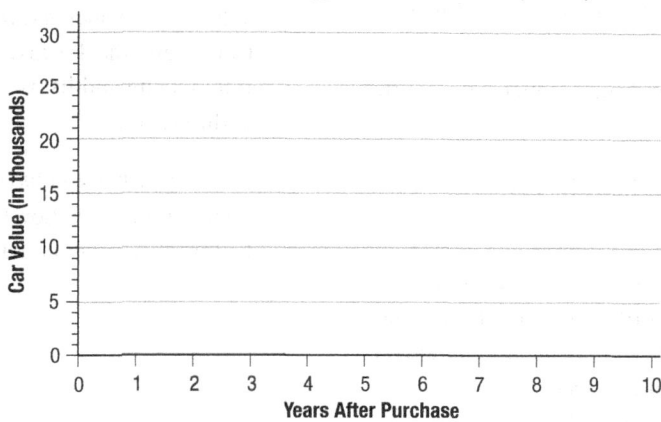

Manufacturer's Suggested Retail Price (MSRP)

(Graph: x-axis labeled "Years After Purchase" from 0 to 10; y-axis labeled "Car Value (in thousands)" from 0 to 30.)

1. The x-axis of the graph is labeled "Years after Purchase." Recall that the MSRP value for each car is for the year 2022 when the car was purchased.

    a. What value on the x-axis will correspond to the year 2022?

    b. Using the value from part a. as the x-coordinate and the MSRP values in column two as the y-coordinates, plot three points on the graph corresponding to the value of the three cars at time of purchase.

    c. What value on the x-axis will correspond to the year 2025?

    d. Using the value from part c. as the x-coordinate and the expected car values in column three as the y-coordinates, plot three points on the graph corresponding to the value of the three cars in 2025.

2. Draw a line segment on the graph connecting the pair of points for each car model. Label each line segment after the car model it represents and label each point with a coordinate pair, (x, y). Consider using a different color when plotting each line segment to help you identify the three models.

3. Use the slope formula, $m = \dfrac{y_2 - y_1}{x_2 - x_1}$, to answer the following questions.

   a. Calculate the rate of depreciation for each model by calculating the slope (or rate of change) between each pair of corresponding points using the slope formula and enter it into the appropriate row of column four of the table.

   b. Are the slopes calculated above positive or negative? Explain why.

   c. Interpret the meaning of the slope for the Toyota Camry making sure to include the units for the variables.

   d. Which car model depreciates in value the fastest? Explain how you determined this.

4. Use the slope-intercept form of an equation, $y = mx + b$, for the following problems.

   a. Write an equation to model the depreciation in value over time of each car (in years). Place these in column five of the table.

   b. What does the y-intercept represent for each car?

5. Use the equations from Problem 4 for the following problems.

   a. Predict the value of the Mini Cooper 4 years after purchase.

   b. Predict the value of the Ford Taurus $2\dfrac{1}{2}$ years after purchase.

6. Determine from the graph how long it takes from the time of purchase until the Ford Taurus and the Toyota Camry have the same value. (It may be difficult to read the coordinates for the point of intersection, but you can get a rough idea of the value from the graph. You can find the exact point of intersection by setting the two equations equal to one another and solving for x.)

   a. After how many years are the car values for the Ford Taurus and the Toyota Camry the same? Round to the nearest tenth.

   b. What is the approximate value of both cars at this point in time? Round to the nearest 100 dollars.

7. How long will it take for the Toyota Camry to fully depreciate (reach a value of zero)?

   a. For the first method, extend the line segment between the two points plotted for the Toyota Camry until it intersects the horizontal axis. The x-intercept is the time at which the value of the car is zero.

   b. Substitute 0 for y in the equation you developed for the Toyota Camry and solve for x. Round to the nearest year.

   c. Compare the results from parts a. and b. Are the results similar? Why or why not?

8. How long will it take for the Ford Taurus to fully depreciate? (Repeat Problem 7 for the Ford Taurus.) Round to the nearest year.

9. Why is there such a difference in depreciation for the Camry and the Taurus? Do some research on a reliable Internet site and list two reasons why cars depreciate at different rates.

10. Based on what you have learned from this activity, do you think retention value will be a significant factor when you purchase your next car? Why or why not?

Chapter 10 Project

Demand and It Shall Be Supplied

An activity to demonstrate the use of linear equations and linear inequalities in real life.

In economics, the demand for a product is the number of units of the product that the market is willing to absorb at a certain price. That is, the demand is the number of units of the product that sells at any given time. The most basic model for the demand d as a function of the price p is given by a linear equation

$$d = b + mp,$$

where m and b are real numbers.

1. The owner of a T-shirt company believes that the demand d (in units) for their Stranger Things T-shirt follows the linear demand model from the introduction, where p is the price of one T-shirt and m and b are real numbers. The company owner knows that they can sell 300 Stranger Things T-shirts for $20 each but only 250 T-shirts at $25 each.

   a. Explain why it is reasonable to believe that the demand of T-shirts decreases as the price increases.

   b. Compute the value of the real number m. How would you interpret the value you have found?

   c. Compute the value of the real number b. How would you interpret the value you have found?

   d. Write the demand equation using your answers from parts b. and c.

The supply for a product is the number of units of a product that the manufacturer can make available at a certain price. The most basic model for the supply s as a function of the price p is given by a linear equation

$$s = b + mp,$$

where m and b are real numbers.

2. The T-shirt company from Problem 1 can produce 275 shirts when the price is $20. If the price is raised to $25, they can produce 300 shirts.

   a. Explain why it is reasonable to believe that the supply of T-shirts increases as the price increases.

   b. Compute the value of the real number m. How would you interpret the value you have found?

   c. Compute the value of the real number b. How would you interpret the value you have found?

   d. Write the supply equation using your answers from parts b. and c.

3. The equilibrium price is the price for which the demand is equal to the supply. At this price, the number of units produced is exactly the number of units absorbed by the market.

   a. Graph the demand and supply equations for the T-shirt company on the same coordinate plane. Recall that the x-axis should represent price.

   b. Find the equilibrium price rounded to the nearest cent and explain what it means in words.

4. Some products have a demand that is not sensitive to changes in price. That is, a large variation in price will not produce a corresponding large variation in demand. This phenomenon is defined as inelastic demand. Perform an internet search to find an example of a product with an inelastic demand. Explain why the demand for the product is inelastic.

Math@Work

Basic Inventory Management 305

Hospitality Management: Preparing for a Dinner Service. 307

Bookkeeper 309

Pediatric Nurse 311

Architecture 313

Statistician: Quality Control 315

Dental Assistant 317

Financial Advisor 319

Market Research Analyst 321

Chemistry 323

Astronomy 325

Math Education 327

Physics 329

Forensic Scientist 331

Other Careers in Mathematics 333

Math Knowledge Required for Math@Work Career Explorations

The following table summarizes the math knowledge required for each Math@Work career exploration project. Use this table to determine when you are ready to explore each career.

Math@Work Project Skill Requirements

Math@Work Project Focus	Whole Numbers	Fractions	Integers	Decimal Numbers	Averages	Percents	Simple Interest	Ratios	Proportions	Geometry	Statistics	Graphing	Linear Equations	Systems of Equations	Mixture Problems	Scientific Notation	Greatest Common Factor	Rational Expressions	Radicals
Basic Inventory Management	✓																		
Hospitality Management	✓	✓			✓														
Bookkeeper				✓															
Pediatric Nurse				✓				✓	✓										
Architecture				✓						✓									
Statistician: Quality Control				✓							✓	✓							
Dental Assistant				✓		✓													
Financial Advisor				✓		✓	✓						✓						
Market Research Analyst				✓									✓						
Chemistry				✓											✓	✓			
Astronomy				✓												✓			
Math Education			✓														✓		
Physics		✓																✓	
Forensic Scientist				✓		✓													✓
Other Careers in Mathematics																			

Math@Work
Basic Inventory Management

As a business manager, you will need to evaluate the company's inventory several times per year. While evaluating the inventory, you will need to ensure that enough of each product will be in stock for future sales based on current inventory count, predicted sales, and product cost. Let's say that you check the inventory four times a year, or quarterly. You will be working with several people to get all of the information you need to make the proper decisions. You need the sales team to give you accurate predictions of how much product they expect to sell. You need the warehouse manager to keep an accurate count of how much of each product is currently in stock and how much of that stock has already been sold. You will also have to work with the product manufacturer to determine the cost to produce and ship the product to your company's warehouse. It's your job to look at this information, compare it, and decide what steps to take to make sure you have enough of each product in stock for sales needs. A wrong decision can potentially cost your company a lot of money.

Suppose you get the following reports: an inventory report of unsold products from the warehouse manager and the report on predicted sales for the next quarter (three months) from the sales team.

Unsold Products	
Item	Number in Stock
A	5025
B	150
C	975
D	2000

Predicted Sales	
Item	Expected Sales
A	4500
B	1625
C	1775
D	2150

Suppose the manufacturer gives you the following cost list for the production and shipment of different amounts of each inventory item.

Item	Amount	Cost	Amount	Cost	Amount	Cost
A	500	$875	1000	$1500	1500	$1875
B	500	$1500	1000	$2500	1500	$3375
C	500	$250	1000	$400	1500	$525
D	500	$2500	1000	$4250	1500	$5575

1. Which items and how much of each item do you need to purchase to make sure the inventory will cover the predicted sales?

2. If you purchase the amounts from Problem 1, how much will this cost the company?

3. By ordering the quantities you just calculated, you are ordering the minimum of each item to cover the expected sales. If the actual sales during the quarter are higher than expected, what might happen? How would you handle this situation?

4. Which math skills were necessary to help you make your decisions?

Math@Work

Hospitality Management: Preparing for a Dinner Service

As the manager of a restaurant, you will need to make sure everything is in place for each meal service. This means that you need to predict and prepare for busy times, such as a Friday night dinner rush. To do this, you will need to obtain and analyze information to determine how much of each meal is typically ordered. After you estimate the number of meals that will be sold, you need to communicate to the chefs how much of each item they need to expect to prepare. An additional aspect of the job is to work with the kitchen staff to make sure you have enough ingredients in stock to last throughout the meal service.

You are given the following data, which are the sales records for the signature dishes during the previous four Friday night dinner services.

Week	Meal A	Meal B	Meal C	Meal D
1	30	42	28	20
2	35	38	30	26
3	32	34	26	26
4	30	32	28	22

Meal C is served with a risotto, a type of creamy rice. The chefs use the following recipe, which makes 6 servings of risotto, when they prepare Meal C. (**Note:** The abbreviation for tablespoon is T and the abbreviation for cup is c.)

$5\frac{1}{2}$ c chicken stock $2\frac{1}{3}$ T chopped shallots $\frac{1}{2}$ c red wine

$1\frac{1}{2}$ c rice   2 T chopped parsley $4\frac{3}{4}$ c thinly sliced mushrooms

2 T butter   2 T olive oil $\frac{1}{2}$ c Parmesan cheese

1. For the past four Friday night dinner services, what was the average number of each signature meal served? If the average isn't a whole number, explain why you would round this number either up or down.

2. Based on the average you obtained for Meal C, calculate how much of each ingredient your chefs will need to make the predicted amount of risotto.

3. The head chef reports the following partial inventory: $10\frac{3}{4}$ c rice, $15\frac{3}{4}$ c mushrooms, and 10 T shallots. Do you have enough of these three items in stock to prepare the predicted number of servings of risotto?

4. Which math skills helped you make your decisions?

Math@Work
Bookkeeper

As a bookkeeper, you will often receive bills and receipts for various purchases or expenses from employees of the company you work for. You will need to split the bill by expense code, assign costs according to customer, and reimburse an employee for their out-of-pocket spending. To do this you will need to know the company's reimbursement policies, the expense codes for different spending categories, and which costs fall into a particular expense category.

Suppose two employees from the sales department recently completed sales trips. Employee 1 flew out of state and visited two customers, Customer A and Customer B. This employee had a preapproved business meal with Customer B and was traveling for three days. Employee 2 drove out of state to visit Customer C. This employee stayed at a hotel for the night and then drove back the next day. The expenses for the two employees are as follows.

Employee 1	
Flight and Rental Car	$470.50
Hotel	$278.88
Meals	$110.56
Business Meal	$102.73
Presentation Materials	$54.86

Employee 2	
Miles Driven	578.5 miles
Fuel	$61.35
Hotel	$79.60
Meals	$53.23
Presentation Materials	$67.84

The expense categories used by your company to track spending are: Travel (includes hotel, flights, mileage, etc.), Meals (business), Meals (travel), and Supplies. Traveling employees are reimbursed up to $35 per day for meals while traveling and for all preapproved business meals. They also receive $0.565 per mile driven with their own car, in addition to the amount they spend on fuel.

1. How much will you reimburse each employee for travel meals? Did either employee go over their allowed meal reimbursement amount?

2. What were the total expense amounts reimbursed for each employee?

3. The company you work for keeps track of how much is spent on each customer. When a salesperson visits multiple customers during one trip, the tracked costs are split between the customers. Fill in this table according to how much was spent on each customer for the different expense categories. (**Note:** For meals, only include the amount the employee was reimbursed.)

Expense	Customer A	Customer B	Customer C
Travel			
Meals (business)			
Meals (travel)			
Supplies			
Total			

Math@Work
Pediatric Nurse

As a pediatric nurse working in a hospital setting, you will be responsible for taking care of several patients during your workday. You will need to administer medications, set IVs, and check each patient's vital signs (such as temperature and blood pressure). While doctors prescribe the medications that nurses need to administer, it is important for nurses to double–check the dosage amounts. Administering the incorrect amount of medication can be detrimental to the patient's health.

During your morning nursing round, you check in on three new male patients and obtain the following information.

	Patient A	Patient B	Patient C
Age	10	9	12
Weight (pounds)	81	68.5	112
Blood Pressure	97/58	100/59	116/73
Temperature (°F)	99.7	97.3	101.4
Medication	A	B	A

The following table shows the bottom of the range for abnormal blood pressure (BP) for boys. If either the numerator or the denominator of the blood pressure ratio is greater than or equal to the values in the chart, this can indicate a stage of hypertension.

Abnormal Blood Pressure for Boys by Age	
	Systolic BP / Diastolic BP
Age 9	109/72
Age 10	111/73
Age 11	113/74
Age 12	115/74

Source: http://www.nhlbi.nih.gov/health/public/heart/hbp/bp_child_pocket/bp_child_pocket.pdf

Medication Directions	
Medication	Dosage Rate
A	40 mg per 10 pounds
B	55 mg per 10 pounds

1. Do any of the patients have a blood pressure that may indicate they have hypertension? If yes, which patient(s)?

2. Use proportions to determine the amount of medication that should be administered to each patient based on weight. Round to the nearest 10 pounds before calculating.

3. The average body temperature is 98.2 degrees Fahrenheit. You are supposed to alert the doctor on duty if any of the patients have a temperature 2.5 degrees higher than average. For which patients would you alert a doctor?

4. Which math skills were necessary to help you make your decisions?

Math@Work
Architecture

As a project architect, you will be part of a team that creates detailed drawings of the project that will be used during the construction phase. It will be your job to ensure that the project will meet guidelines given to you by your company, such as square footage requirements and budget constraints. You will also need to meet the design requirements requested by the client.

Suppose you are part of a team that is designing an apartment building. You are given the task to create the floor plan for an apartment unit with two bedrooms and one bathroom. The apartment management company that has contracted your company to do the project has several requirements for this specific apartment unit.

1. One bedroom is the "master bedroom" and must have at least 60 square feet more than the other bedroom.
2. All walls must intersect or touch at 90 degree angles.
3. The kitchen must have an area of no more than 110 square feet.
4. The apartment must be between 1000 square feet and 1050 square feet.

A preliminary sketch of the apartment is shown here.

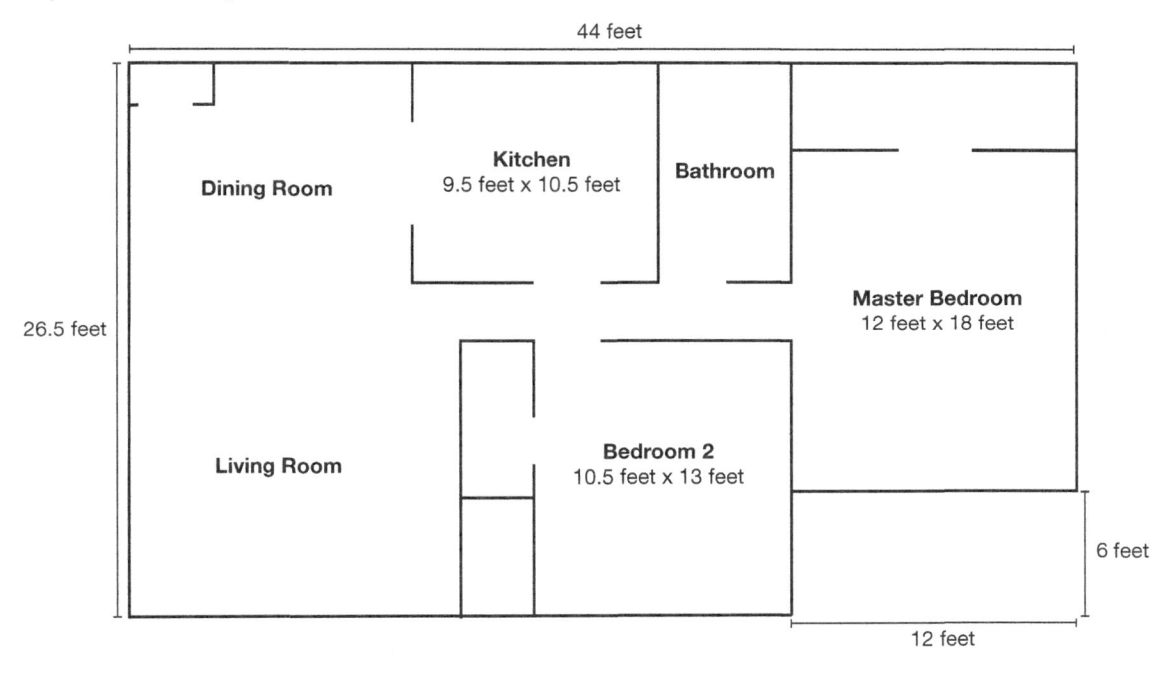

1. Does the apartment have the required total square footage that was requested? Is it over or under the total required?

2. Does the apartment blueprint meet the other requirements given by the client? If not, what does not meet the requirements?

3. For this specific apartment unit, the total construction cost per square foot is estimated to be $75.75. Approximately how much will it cost to construct each two-bedroom apartment based on the floor plan?

Name: Date: **369**

Math@Work
Statistician: Quality Control

Suppose you are a statistician working in the quality control department of a company that manufactures the hardware sold in kits to assemble bookshelves, TV stands, and other ready-to-assemble furniture pieces. There are three machines that produce a particular screw and each machine is sampled every hour. A measurement of the screw length is determined with a micrometer, which is a device used to make highly precise measurements. The screw is supposed to be 3 inches in length and can vary from this measurement by no more than 0.1 inches or it will not fit properly into the furniture. The following table shows the screw length measurements (in inches) taken each hour from each machine throughout the day. The screw length data from each machine has also been plotted

Screw Length Measurements (in inches)			
Sample Time	Machine A	Machine B	Machine C
8 a.m.	2.98	2.92	2.99
9 a.m.	3.00	2.94	3.00
10 a.m.	3.02	2.97	3.01
11 a.m.	2.99	2.96	3.03
12 p.m.	3.01	2.94	3.05
1 p.m.	3.00	2.95	3.04
2 p.m.	2.97	2.93	3.06
3 p.m.	2.99	2.92	3.08
Mean			
Range			

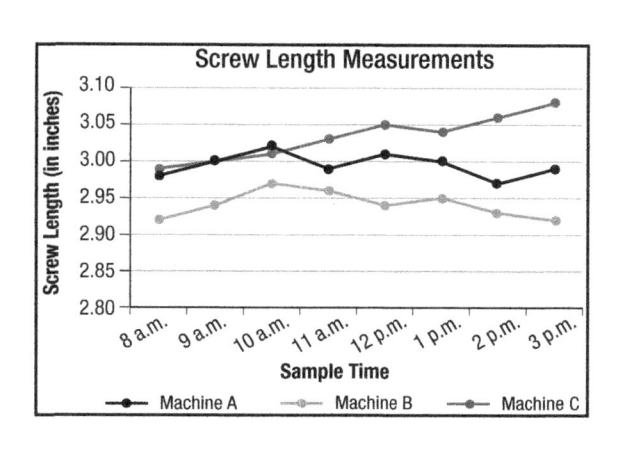

1. Calculate the mean and range of the data for each machine and place them in the bottom two rows of the table.

2. If the screw length can vary from 3 inches by no more than 0.1 inches (plus or minus), what are the lowest and highest values for length that will be acceptable? Place a horizontal line on the graph at each of these values on the vertical axis. These are the tolerance or specification limits for screw length.

3. Have any of the three machines produced an unacceptable part today? Are any of the machines close to making a bad part? If so, which one(s)?

4. Look at the graph and the means from the table that show the average screw length produced by each machine. Draw a bold horizontal line on the graph at 3 to emphasize the target length. Do all the machines appear to be making parts that vary randomly around the target of 3 inches?

5. Look at the range values from the table. Do any of the machines appear to have more variability in the length measurements than the others?

6. In your opinion, which machine is performing best? Would you recommend that any adjustments be made to any of the machines? If so, which one(s) and why?

Math@Work
Dental Assistant

As a dental assistant, your job duties will vary depending on where you work. Suppose you work in a dental office where you assist with dental procedures and managing patients' accounts. When a patient arrives for their appointment, you will need to review their chart and make sure they are up to date on preventive care, such as X-rays and cleanings. When the patient leaves, you will need to fill out an invoice to determine how much to charge the patient for their visit.

Dental patients generally have a new X-ray taken yearly. Cleanings are performed every 6 months, although some patients have their teeth cleaned more often. The following table shows the date of the last X-ray and cleaning for three patients that are visiting the office today. (**Note:** All dates are within the past year.)

Patient Histories		
Patient	**Last X–ray**	**Last Cleaning**
A	April 15	October 20
B	June 6	January 12
C	October 27	October 27

During Patient A's visit, she received a fluoride treatment and a cleaning. Patient A has no dental insurance. During Patient B's visit, he received a filling on one surface of a tooth. Patient B has dental insurance which pays for 60% of the cost of fillings. During Patient C's visit, he had a cleaning, a filling on one surface of a tooth, and a filling on two surfaces of another tooth. Patient C has dental insurance which covers the full cost of cleanings and 50% of the cost of fillings.

Fee Schedule	
Procedure	**Cost**
Cleaning	$95
Fluoride treatment	$35
Filling, One surface	$175
Filling, Two surfaces	$235
X–ray, Panoramic	$110

1. Using today's date, determine which of the three patients are due for a dental cleaning in the next two months?

2. Using today's date, determine which of the patients will require a new set of X-rays during this visit.

3. Determine the amount each patient will be charged for their visit (without insurance). Don't forget to include the cost of any X-rays that are due during the visit.

4. Use the insurance information to determine the amount that each patient will pay out-of-pocket at the end of their visit.

Math@Work

Financial Advisor

As a financial advisor working with a new client, you must first determine how much money your client has to invest. The client may have a lump sum that they have saved or inherited, or they may wish to contribute an amount monthly from their current salary. In the latter case, you must then have the client do a detailed budget so that you can determine a reasonable amount that the client can afford to set aside on a monthly basis for investment.

The second piece of information necessary when dealing with a new client is determining how much risk-tolerance they have. If the client is young or has a lot of money to invest, they may be willing to take more risk and invest in more aggressive, higher interest-earning funds. If the client is older and close to retirement or has little money to invest, they may prefer less-aggressive investments where they are essentially guaranteed a certain rate of return. The range of possible investments that would suit each client's needs and goals are determined using a survey of risk-tolerance.

Suppose you have a client who has a total of $25,000 to invest. You determine that there are two investment funds that meet the client's investment preferences. One option is an aggressive fund that earns an average of 12% interest and the other is a more moderate fund that earns an average of 5% interest. The client desires to earn $2300 this year from these investments.

Investment Type	Principal Invested	·	Interest Rate	=	Interest Earned
Aggressive Fund	x				
Moderate Fund					

To determine the amount of interest earned you know to use the table above and the formula $I = Prt$, where I is the interest earned, P is the principal or amount invested, r is the average rate of return, and t is the length of time invested. Since the initial investment will last one year, $t = 1$.

1. Fill in the Principal Invested and Interest Rate columns of the table with the known information about the principal invested. If x is the amount invested in the aggressive fund and the total amount to be invested is $25,000, create an expression involving x for the amount that will be left to invest in the moderate fund. Place this expression in the appropriate cell of the table.

2. Determine an expression in x for the interest earned on each investment type by multiplying the principal by the interest rate.

3. Determine the amount invested in each fund by setting up an equation using the expressions in the Interest Earned column and the fact that the client desires to earn $2300 from the interest earned on both investments.

4. Verify that the investment amounts calculated for each fund in the previous step are correct by calculating the actual interest earned in a year for each and making sure they sum to $2300.

5. Why would you not advise your client to invest all their money in the fund earning 12% interest, even though it has the highest average interest rate?

Name: Date: **375**

Math@Work
Market Research Analyst

As a market research analyst, you may work alone at a computer, collecting and analyzing data, and preparing reports. You may also work as part of a team or work directly with the public to collect information and data. Either way, a market research analyst must have strong math and analytical skills and be very detail-oriented. They must have strong critical-thinking skills to assess large amounts of information and be able to develop a marketing strategy for the company. They must also possess good communication skills in order to interpret their research findings and be able to present their results to clients.

Suppose you work for a shoe manufacturer who wants to produce a new type of lightweight basketball sneaker similar to a product a competitor recently released into the market. You have gathered some sales data on the competitor in order to determine if this venture would be worthwhile, which is shown in the table below. To begin your analysis, you create a scatter plot of the data to see the sales trend. (A scatter plot is a graph made by plotting ordered pairs in a coordinate plane in order to show the relationship between two variables.) You determine that the x-axis will represent the number of weeks after the competitor's new sneaker went on the market and the y-axis will represent the amount of sales in thousands of dollars.

Number of Weeks x	Sales (in 1000s) y
3	15
6	22
9	28
12	35
15	43

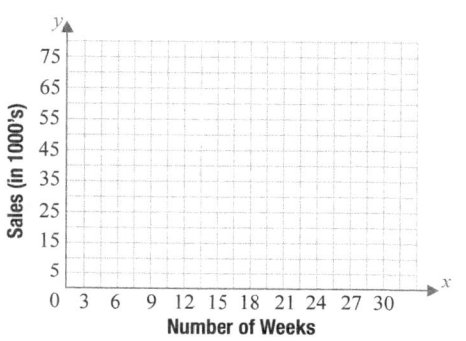

1. Create a scatter plot of the sales data by plotting the ordered pairs in the table on the coordinate plane. Does the data on the graph appear to follow a linear pattern? If so, sketch a line that you feel would "best" fit this set of data. (A market research analyst would typically use computer software to perform a technique called regression analysis to fit a "best" line to this data.)

2. Using the ordered pairs corresponding to weeks 9 and 15, find the equation of a line running through these two data points.

3. Interpret the value calculated for the slope of the equation in Problem 2 as a rate of change in the context of the problem. Write a complete sentence.

4. If you assume that the sales trend in sneaker sales follows the model determined by the linear equation in Problem 2, predict the sneaker sales in 6 months. Use the approximation that 1 month is equal to 4 weeks.

5. Give at least two reasons why the assumption made in Problem 4 may be invalid.

Math@Work
Chemistry

As a pharmaceutical chemist, you will need an advanced degree in pharmaceutical chemistry, which combines biology, biochemistry, and pharmaceuticals. In this career, you will most likely spend your day in a lab setting creating new medications or researching their effectiveness. You will often work as part of a team working towards a joint goal. As a result, in addition to strong math skills and an understanding of chemistry, you will need to have good communication and leadership skills. Since you will be working directly with chemicals, you will also need to have a strong understanding of lab safety rules to ensure the safety of not only yourself but your coworkers as well.

Suppose you work at a pharmaceutical company which creates and produces medications for various skin conditions. You are currently on a team which is developing an acne-controlling facial cleanser. Your team is working on determining the gentlest formula possible that is still effective so that the cleanser can be used on sensitive skin. Half of your team is working with salicylic acid and the other half is working with benzoyl peroxide.

As a part of your work, you will need to keep up on current research. Learning about new chemicals, new methods, and new research will be a continuous part of your life.

1. Perform an Internet search for benzoyl peroxide. How does it work to clean skin and prevent acne?

2. Perform an Internet search for salicylic acid. How does it work to clean skin and prevent acne?

3. Based on your research, which chemical seems better suited to treat acne on sensitive skin?

Another aspect of your career will involve the mixing of chemicals to create new compounds. Having the correct concentrations of chemicals is also important so the resulting solution works as you expect it to. When you don't have the correct concentration of a chemical in stock, it is possible to mix two concentrations together to obtain the desired concentration.

4. Your team wants to create a cleanser with 4% benzoyl peroxide. The lab currently has 2.5% and 10% concentrations of benzoyl peroxide in stock. To create 500 mL of 4% benzoyl peroxide, how much of each concentration should be combined?

Math@Work
Astronomy

Astronomy is the study of celestial bodies, such as planets, asteroids, and stars. While you work in the field of astronomy, you will use knowledge and skills from several other fields, such as mathematics, physics, and chemistry. An important tool of astronomers is the telescope. Several powerful telescopes are housed in observatories around the world. One of the many things astronomers use observatories for is discovering new celestial objects such as a near-Earth object (NEO). NEOs are comets, asteroids, and meteoroids that orbit the sun and cross the orbital path of Earth. The danger presented by NEOs is that they may strike the Earth and result in global catastrophic damage. (**Note:** The National Aeronautics and Space Administration (NASA) keeps track of all NEOs which are a potential threat at the website cneos.jpl.nasa.gov/sentry.)

For an asteroid to be classified as an NEO, the asteroid must have an orbit that partially lies within 0.983 and 1.3 astronomical units (AU) from the sun, where 1 AU is the furthest distance from the Earth to the sun, approximately 9.3×10^7 miles.

Near-Earth Object Distance			
	Minimum		Maximum
Distance in AU	0.983 AU	1 AU	1.3 AU
Distance in Miles		9.3×10^7 miles	

Suppose you discover three asteroids that you suspect may be NEOs. You perform some calculations and come up with the following facts. The furthest that Asteroid A is ever from the sun is 81,958,000 miles. The closest Asteroid B is ever to the sun is 125,290,000 miles. The closest Asteroid C is ever to the sun is 92,595,000 miles.

1. To determine if any of the asteroids pass within the range to be classified as an NEO, fill in the missing values from the table.

2. Based on the measurements from Problem 1, do any of the three asteroids qualify as an NEO?

There are two scales that astronomers use to explain the potential danger of NEOs. The Torino Scale is a scale from 0 to 10 that indicates the chance that an object will collide with the Earth. A rating of 0 means there is an extremely small chance of a collision and a 10 indicates that a collision is certain to happen. The Palermo Technical Impact Hazard Scale is used to rate the potential impact hazard of an NEO. If the rating is less than −2, the object poses a very minor threat with no drastic consequences if the object hits the Earth. If the rating is between −2 and 0, then the object should be closely monitored as it could cause serious damage.

Go to the NASA website cneos.jpl.nasa.gov/sentry to answer the following questions.

3. Does any NEO have a Torino Scale rating higher than 0? If so, what is the object's designation (or name) and during which year range could a potential impact occur?

4. Which NEO has the highest Palermo Scale rating? During which year range could a potential impact occur?

Name: Date: **381**

Math@Work
Math Education

As a math instructor at a public high school, your day will be spent preparing class lectures, grading assignments and tests, and teaching students with a wide variety of backgrounds. While teaching math, it is your job to explain the concepts and skills of math in a variety of ways to help students learn and understand the material. As a result, a solid understanding of math and strong communication skills are very important. Teaching math is a challenge and being able to understand the reasons that students struggle with math and empathize with these students is a critical aspect of the job.

Suppose that the next topics you plan to teach to your algebra students involve finding the greatest common factor and factoring by grouping. To teach these skills, you will need to plan how much material to cover each day, choose examples to walk through during the lecture, and assign in-class work and homework. You decide to spend the first day on this topic explaining how to find the greatest common factor of a list of integers.

1. It is usually easier to teach a group of students a new topic by initially showing them a single method. If a student has difficulty with that method, then showing the student an alternative method can be helpful. Which method for finding the greatest common factor would you teach to the class during the class lecture?

2. On a separate piece of paper, sketch out a short lecture on finding the greatest common factor of a list of integers. Be sure to include examples that range from easy to difficult.

3. While the class is working on an in-class assignment, you find that a student is having trouble following the method that you taught to the entire class. Describe an alternative method that you could show the student.

4. From your experience with learning how to find the greatest common factor of a list of integers, what do you think are some areas that might confuse students and cause them to struggle while learning this topic? Explain how understanding the areas that might cause confusion can help you become a better teacher.

Math@Work

Physics

As an employee of a company that creates circuit boards, your job may vary from designing new circuit boards, setting up machines to mass–produce the circuit boards, to testing the finished circuit boards as part of quality control. Depending on your position, you may work alone or as part of a team. Regardless of who you work with, you will need strong math skills to be able to create new circuit board designs and strong communication skills to describe the specifications for a new circuit board design, describe how to set up the production line, or explain why a part is faulty.

Suppose your job requires you to create new circuit boards for a variety of electronic equipment. The latest circuit board that you are designing is a small part of a complicated device. The circuit board you create has three resistors which run in parallel, as shown in the diagram.

Two of the resistors were properly labeled with their correct resistance, which is measured in ohms. The first resistor has a rating of 2 ohms. The second resistor has a rating of 3 ohms. The third resistor was taken from the supply shelf for resistors of a certain rating, but the resistor was unlabeled. As a result, you are unsure if it has the correct resistance for the current you want to produce. You use an ohmmeter, a device that measures resistance in a circuit, to determine that the total resistance of the circuit you created is $\frac{30}{31}$ ohms.

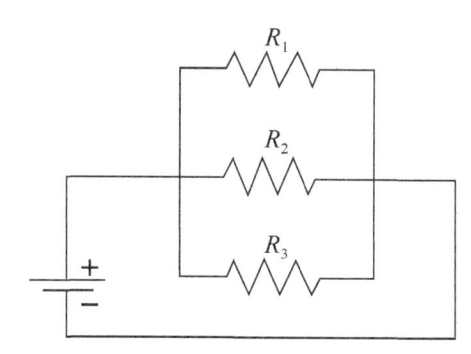

You know that the equation to determine the total resistance R_t is $\frac{1}{R_t} = \frac{1}{R_1} + \frac{1}{R_2} + \frac{1}{R_3}$, where R_1 is the resistance of the first resistor, R_2 is the resistance of the second resistor, and R_3 is the resistance of the third resistor.

1. Use the formula to determine the resistance of the third resistor given that the total resistance of the circuit is $\frac{30}{31}$ ohms.

2. Was the third resistor on the correct shelf if you took it from the supply shelf that holds resistors with a rating of 7 ohms?

3. What would be the total resistance of the circuit if the third resistor had a rating of 7 ohms?

4. What do you think would happen if the resistance of the unlabeled resistor wasn't determined and the circuit board was sent to the production line to be mass–produced?

Math@Work

Forensic Scientist

As a forensic scientist, you will work as part of a team to investigate the evidence from a crime scene. Every case you encounter will be unique and the work may be intense. Communication is especially important because you will need to be clear and honest about your findings and your conclusions. A suspect's freedom may depend on the conclusions your team draws from the evidence.

Suppose the most recent case that you are involved in is a hit-and-run accident. A body was found at the side of the road with skid marks nearby. The police are unsure if the cause of death of the victim was vehicular homicide. Among the case description, the following information is provided to you.

Accident Report	
Date:	June 14
Time:	9:30 pm
Climate:	55 degrees Fahrenheit, partly cloudy, dry
Description of crime scene:	
Victim was found at the side of a road. Body temperature upon arrival is 84.9 °F. Posted speed limit is 30 mph. Road is concrete. Conditions are dry. Skid marks near the body are 88 feet in length.	

Known formulas and data:

A body will cool at a rate of 2.7 °F per hour until the body temperature matches the temperature of the environment. Average human body temperature is 98.6 °F.

Impact Speed and Risk of Death	
Impact Speed	**Risk of Death**
23 mph	10%
32 mph	25%
42 mph	50%
58 mph	90%
Source: 2011 AAA Foundation for Traffic Safety "Impact Speed and Pedestrian's Risk of Severe Injury or Death"	

Braking distance is calculated using the formula $\frac{s}{\sqrt{l}} = k$, where s is the initial speed of the vehicle in mph, l is the length of the skid marks in feet, and k is a constant that depends on driving conditions. Based on the driving conditions on that road for the last 12 hours, $k = \sqrt{20}$.

1. Based on the length of the skid marks, how fast was the car traveling before it attempted to stop? Round to the nearest whole number.

2. Based on the table, what percent of pedestrians die after being hit by a car moving at that speed?

3. Based on the cooling of the body, if the victim died instantly, how long ago did the accident occur? Round to the nearest hour.

4. Can you think of any other factors that should be taken into consideration before determining whether the impact of the car was the cause of death?

Name: Date:

Math@Work
Other Careers in Mathematics

Earning a degree in mathematics or minoring in mathematics can open many career pathways. While a degree in mathematics or a field which uses a lot of mathematics may seem like a difficult path, it is something anyone can achieve with practice, patience, and persistence. Three growing fields of study which rely on mathematics are actuarial science, computer science, and operations research. While each of these fields involves mathematics, they require special training or additional education outside of a math degree. A brief description of each career is provided below along with a source to find more information about these careers.

Growing Fields of Study

Actuarial Science: The field of actuarial science uses methods of mathematics and statistics to evaluate risk in industries such as finance and insurance. Visit www.beanactuary.org for more information

Computer Science: From creating web pages and computer programs to designing artificial intelligence, computer science uses a variety of mathematics. Visit www.acm.org for more information.

Operations Research: The discipline of operations research uses techniques from mathematical modeling, statistical analysis, and mathematical optimization to make better decisions, such as maximizing revenue or minimizing costs for a business. Visit www.informs.org for more information.

There are numerous careers that have not been discussed in this workbook. Exploring career options before choosing a major is a very important step in your academic career. Learning about the career you are interested in before completing your degree can help you choose courses that will align with your career goals. You should also explore the availability of jobs in your chosen career and whether you will have to relocate to another area to be hired. The following web sites will help you find information related to different careers that use mathematics. Another great resource is the mathematics department at your college.

The **Mathematical Association of America** has a website with information about several careers in mathematics. Visit www.maa.org/careers to learn more.

The **Society for Industrial and Applied Mathematics** also has a webpage dedicated to careers in mathematics. Visit www.siam.org/careers to learn more.

The **Occupational Outlook Handbook** is a good source for information on educational requirements, salary ranges, and employability of many careers, not just those that involve mathematics. Visit http://www.bls.gov/ooh/ to learn more.

Answer Key

Chapter 1: Whole Numbers

1.1 Exercises

Concept Check
1. True
3. True

Practice
5. 2: ten thousands, 4: thousands, 6: hundreds, 8: ones
7. Six hundred eighty-three thousand, one hundred

Applications
9. Eighty-two thousand, one hundred three

Writing & Thinking
11. 0 is a whole number and not a natural number. Both whole numbers and natural numbers include 1, 2, 3, 4, 5, and so on.

1.2 Exercises

Concept Check
1. False; A polygon has three or more sides.
3. False; Borrowing must occur.

Practice
5. 58
7. 144
9. 42 cm

Applications
11. $39,100

Writing & Thinking
13. Answers will vary. Subtraction may be used when paying bills, buying items, losing weight, etc.

1.3 Exercises

Concept Check
1. False; The numbers being multiplied are called factors.
3. True

Practice
5. 2352
7. Associative property of multiplication
9. 40 square centimeters

Applications
11. 8928 slices of bread

Writing & Thinking
13. Because any number multiplied by 1 results in the original number.

1.4 Exercises

Concept Check
1. False; If a division problem has a zero remainder...
3. False; 12 ÷ 0 is undefined.

Practice
5. 0
7. 9

Applications
9. 16 grams

Writing & Thinking
11. To check a division problem, multiply the quotient and divisor, and then add the remainder. The result should equal the original dividend.

1.5 Exercises

Concept Check
1. True
3. True

Practice
5. 220; 223
7. 40,000; 43,680

Applications
9. $40,000; $35,316

Writing & Thinking
11. Estimation uses rounded values to find an approximate sum, difference, product, etc. Answers will vary.

1.6 Exercises

Concept Check
1. True
3. False; Quotient indicates division.

Applications
5. 1103 calories
7. 380 sq in.

Writing & Thinking
9. Answers will vary.

1.7 Exercises

Concept Check
1. False; Equals 81
3. False; 7^0 is 1.

Practice
5. **a.** 2 **b.** 3 **c.** 8
7. 2

Applications
9. **a.** No. Here it shows that we are only dividing the old trading cards by 6 friends versus both the old and new trading cards by 6 friends.
b. 522; $\dfrac{15 \cdot 10 \cdot 20 + 132}{6}$

Writing & Thinking
11. If addition is within parentheses (or other grouping symbols), addition would be performed first.

1.8 Exercises

Concept Check
1. True
3. False; 7605 is divisible by 5.

Practice
5. 3, 5
7. None

Applications
9. 5 people would raise $2480 each; 10 people would raise $1240 each.

Writing & Thinking
11. **a.** 30, 45; Answers will vary.
b. 9, 12; Answers will vary.
c. 10, 25; Answers will vary.

1.9 Exercises

Concept Check
1. False; A prime number has exactly 2 factors.
3. False; 231 is a composite number.

Practice
5. Prime
7. 5^3

Applications
9. 1, 2, 3, 4, 6, 8, 12, 24

Writing & Thinking

11. No, some odd numbers are the product of two or more odd prime factors, for example, $3 \cdot 3 = 9$, $3 \cdot 5 = 15$, $3 \cdot 7 = 21$, etc.

Chapter 2: Fractions and Mixed Numbers

2.1 Exercises

Concept Check

1. False; The numerator is 11.
3. True

Practice

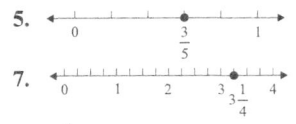

5. (number line with point at $\frac{3}{5}$)
7. (number line with point at $3\frac{1}{4}$)
9. $1\frac{1}{3}$

Applications

11. $\frac{115}{146}$

Writing & Thinking

13. Multiply the denominator by the whole number and add the numerator. This number is the new numerator. The denominator stays the same.

2.2 Exercises

Concept Check

1. True
3. False; The statement $\frac{1}{3} \cdot \frac{2}{5} = \frac{2}{5} \cdot \frac{1}{3}$ is an example of the commutative property of multiplication.

Practice

5. $\frac{1}{4}$
7. $\frac{1}{3}$

Applications

9. $\frac{3}{8}$

Writing & Thinking

11. No. If a fraction is less than 1 then its product with another number will be less than that other number. So, if the other number is less than 1, the product will be less than 1. Answers will vary.

2.3 Exercises

Concept Check

1. False; The reciprocal of 1 is 1.
3. False; The reciprocal of 12 is $\frac{1}{12}$.

Practice

5. $\frac{8}{9}$
7. Undefined

Applications

9. 200 years

Writing & Thinking

11. $0 = \frac{0}{1}$ and the reciprocal would be $\frac{1}{0}$ but division by 0 is undefined. So 0 has no reciprocal.

2.4 Exercises

Concept Check

1. True
3. False; The mixed number $4\frac{1}{5}$ is equal to $\frac{21}{5}$.

Practice

5. $2\frac{1}{6}$
7. 4

Applications

9. a. $22\frac{1}{2}$ gallons
   b. $45

Writing & Thinking

11. a. More, since $10\frac{1}{2} > 5\frac{7}{10}$
    b. $\frac{35}{19}$ or $1\frac{16}{19}$

2.5 Exercises

Concept Check

1. False; The LCM of 15 and 25 is 75.
3. False; The first five multiples of 4 are 4, 8, 12, 16, and 20.

Practice

5. 30
7. a. LCM = 490
   b. $490 = 14 \cdot 35$
      $= 35 \cdot 14 = 49 \cdot 10$
9. 45

Applications

11. a. 360 pieces
    b. 15 boxes, 10 boxes, and 8 boxes, respectively

Writing & Thinking

13. Multiplying the two numbers together will give the LCM if those two numbers have no common factors. If they have any factors in common, then you would only use that common factor once. Examples will vary.

2.6 Exercises

Concept Check

1. True
3. False; LCD stands for least common denominator.
5. True

Practice

7. $\frac{17}{21}$
9. $\frac{23}{42}$

Applications

11. 1 ounce
13. The LCM finds the least common multiple of a set of numbers. The LCD does the same thing for the set of numbers determined by the denominators.

2.7 Exercises

Concept Check

1. True
3. False; LCDs are required when adding or subtracting mixed numbers.

Practice

5. $7\frac{2}{3}$
7. $4\frac{3}{5}$

9. $8\frac{7}{12}$ hours

Writing & Thinking

11. Fractions should be first in case the fraction being subtracted is larger than the other fraction and 1 needs to be borrowed from the whole number.

2.8 Exercises

Concept Check

1. True

3. True

Practice

5. $\frac{17}{20}$ by $\frac{1}{20}$

7. $\frac{2}{3}$

9. $\frac{22}{45}$

Applications

11. $2\frac{1}{12}$ inches

Chapter 3: Decimal Numbers

3.1 Exercises

Concept Check

1. True
3. False; On a number line, any number to the right of another number is larger than that other number.

Practice

5. 2.57
7. 6.028
9. a. 5
   b. 2
   c. 2, 5, 2
   d. 3.0065

Applications

11. Two and eight-hundred twenty-five ten-thousandths

Writing & Thinking

13. Moving left to right, compare digits with the same place value. When one compared digit is larger, the corresponding number is larger.

3.2 Exercises

Concept Check

1. True
3. False; In subtracting decimal numbers, line up the decimal points and corresponding digits vertically.

Practice

5. 50.085
7. 11.131

Applications

9. a. $94.85
   b. $5.15

Writing & Thinking

11. Decimal numbers need to be aligned vertically so that numbers with the same place value are being added together. If not, then a 60 may be added to a 7 as if it were a 6 being added to a 7, giving 13, not the value of 67 that it should be.

3.3 Exercises

Concept Check

1. False; The decimal points do not need to be aligned vertically when multiplying decimal numbers.
3. True

Practice

5. 0.42
7. 0.04336

Applications

9. $240.90

Writing & Thinking

11. In multiplication with decimal numbers, placement of the decimal point must be considered. Otherwise, multiplication with whole numbers and decimal numbers are the same.

3.4 Exercises

Concept Check

1. True
3. True

Practice

5. 0.99
7. 0.01

Applications

9. $23.83

Writing & Thinking

11. In division with decimal numbers, placement of the decimal point must be considered. Also, with whole numbers a remainder is considered while, with decimal numbers, rounding the quotient is considered. Otherwise, division with whole numbers and decimal numbers are the same.

3.5 Exercises

Concept Check

1. True
3. False; When estimating 16.469 ÷ 3.87, the answer would be 5.
5. False; According to the rules for order of operations, multiplication and division should be performed before addition and subtraction.

Practice

7. 20; 26.08
9. 2; 2.05

Applications

11. a. 39 pounds
    b. 35.43 pounds

Writing & Thinking

13. The two procedures would produce different results. Method **a.** is more accurate since you only round one figure. In method **b.** you multiply two rounded amounts which increases the difference between the approximated amount and the actual amount. Answers will vary.

3.6 Exercises

Concept Check

1. True
3. False; In some cases, fractions can be converted to decimal form without losing accuracy.

Practice

5. $\frac{9}{50}$
7. 6.67
9. 0.7

Applications

11. 17.92 inches

Writing & Thinking

13. For the numerator, write the whole number formed by all the digits of the decimal number, and for the denominator, write the power of 10 that corresponds to the rightmost digit. Reduce the fraction, if possible.

Chapter 4: Ratios, Proportions, and Percents

4.1 Exercises

Concept Check

1. True
3. False; The ratio 8:2 can be reduced to the ratio 4:1.

Practice

5. $\dfrac{9}{14}$
7. $\dfrac{\$2 \text{ profit}}{\$5 \text{ invested}}$
9. 113.7¢/oz; 66.6¢/oz; 12 oz at $7.99

Applications

11. $\dfrac{9}{41}$

Writing & Thinking

13. Numerator

4.2 Exercises

Concept Check

1. False; A proportion is a statement that two ratios are equal.
3. True
5. True

Practice

7. False
9. $B = 7.8$

Applications

11. 180 minutes or 3 hours

4.3 Exercises

Concept Check

1. True
3. False; A decimal number that is between 0.01 and 0.10 is between 1% and 10%.

Practice

5. 20%
7. 2%
9. 0.07

Applications

11. 4%

Writing & Thinking

13. Percent means per centum or per 100. For example, fifty-eight percent means 58 out of 100. Percent can be written as a fraction with 100 in the denominator as in 58/100. The decimal equivalent, 0.58 is read as "fifty-eight hundredths," indicating percent can be written using the hundredths place, another connection.

4.4 Exercises

Concept Check

1. False; Fractions that have denominators other than 100 can be changed to a percent.
3. False; When changing from a percent to a mixed number, the fraction should be reduced.

Practice

5. 75%
7. $1\dfrac{1}{5}$

Applications

9. 85%

Writing & Thinking

11. 100% = 1 so anytime there is a mixed number, which has a value greater than 1, the percentage will be greater than 100%. Proper fractions (numerator is smaller than denominator) have a value less than 1 and therefore the percentage will be less than 100%.

4.5 Exercises

Concept Check

1. True
3. False; In the problem "What is 26% of 720?" the missing number is the amount.

Practice

5. 7.5
7. 15
9. 250

Applications

11. $97,600

Writing & Thinking

13. Proportions would work for mixed numbers because a mixed number can be rewritten as a fraction. The only additional step required would be to change the mixed number to an improper fraction and then solve the proportion as normal.

4.6 Exercises

Concept Check

1. False; In order to solve the equation $0.56 \cdot B = 12$ for the base, B, one would divide 12 by 0.56.
3. False; The solution to the problem "50% of what number is 352?" could be found by solving the equation $0.5 \cdot B = 352$.

Practice

5. 7
7. 42
9. 20

Applications

11. $175,000
13. The amount is the number that is often near the word "is." The base is the number that often follows the word "of." The rate is the

number written either as a fraction or as a decimal number that has not been identified as the amount or the base, and usually appears before the word "of."

4.7 Exercises

Concept Check

1. True
3. True

Applications

5. a. $82.50
   b. $192.50

7. $11,700

Writing & Thinking

9. Sales tax and tips are percentages of some item or service. The percent is the rate, while the cost of the item being purchased is the base. The amount is then the sales tax itself, which is being compared to the base. A sales tax might be 8%, as in "what is 8% of the cost of the item purchased?"

4.8 Exercises

Concept Check

1. True
3. False; Compound interest is earned on the principal and interest earned.

Applications

5. $112.50
7. $189.24

Writing & Thinking

9. The simple interest formula is $I = P \cdot r \cdot t$ where I is interest, P is principal, r is rate, and t is time. Interest is the amount of money paid for the use of money. The principal is the starting amount invested. Rate is the interest rate and should be written as a decimal or fraction. Time is the amount of time, in years, that interest is being earned on the principal. Time can be written as a decimal or fraction. When a decimal is used, it should only be when it is a terminating decimal so that no rounding is required, which could change the value calculated.

Chapter 5: Measurement

5.1 Exercises

Concept Check

1. True
3. True

Practice

5. 8
7. 13

Applications

9. $93.22

Writing & Thinking

11. Colby would need to know that there are 3 feet in a yard and 5280 feet in a mile.

5.2 Exercises

Concept Check

1. False; To change from smaller units to larger units, division must be used.
3. False; In metric units, a square that is 1 centimeter long on each side is said to have an area of 1 square centimeter.

Practice

5. 0.01977
7. 1300; 130 000

Applications

9. 1750 railroad ties

Writing & Thinking

11. Each category of metric units has a base unit. The prefixes determine how many or what fraction of the base unit is being used. For example, the basic unit of length is meter and a millimeter is 1/1000 of a meter, a centimeter is 1/100 of a meter, and a kilometer is 1000 meters.

5.3 Exercises

Concept Check

1. True
3. False; In 1 liter there are 1000 milliliters.
5. True

Practice

7. 6300
9. 2

Applications

11. 20 cups

5.4 Exercises

Concept Check

1. False; Water freezes at 32 degrees Fahrenheit.
3. False; A 5k (km) run is shorter than a 5 mile run.

Practice

5. 77
7. 19.35
9. 72.75

Applications

11. 226.3 km

Chapter 6: Geometry

6.1 Exercises

Concept Check

1. True
3. True
5. True

Practice

7. a. Straight
   b. Right
   c. Acute
   d. Obtuse

9. a. 150°
   b. Yes; ∠2 and ∠3 are supplementary.
   c. ∠1 and ∠3; ∠2 and ∠4

Answer Key 7.1 Exercises

d. ∠1 and ∠2; ∠2 and ∠3; ∠3 and ∠4; ∠1 and ∠4

11. Equilateral

Applications

13. a. $m\angle Z = 80°$
  b. Acute
  c. $\overline{YZ}$
  d. $\overline{XZ}$ and $\overline{XY}$
  e. No, no angle is 90°

6.2 Exercises

Concept Check

1. a. True
  b. False; Not all rectangles have four equal sides.
3. True
5. a. B
  b. C
  c. A
  d. E
  e. D

Practice

7. 18 km
9. 35 ft

Applications

11. a. 65 ft
  b. $156

Writing & Thinking

13. Perimeter is the distance around a figure. Formulas for the perimeter of: triangle ($P = a + b + c$), square ($P = 4s$), rectangle ($P = 2l + 2w$), trapezoid ($P = a + b + c + d$), and parallelogram ($P = 2a + 2b$).

6.3 Exercises

Concept Check

1. False; The ($b + c$) in the trapezoid area formula represents the sum of the lengths of the two parallel bases.
3. False; The area formula for a triangle is $A = \dfrac{1}{2}bh$.

Practice

5. 81 ft^2
7. 48 in.^{2}
9. 36 cm^2

Applications

11. a. 75 cm
  b. 336 cm^2

Writing & Thinking

13. The expression ($b + c$) represents the sum of the two bases in a trapezoid. The bases are the two sides that are parallel in a trapezoid.

6.4 Exercises

Concept Check

1. True
3. False; The length of the radius of a circle is half of the length of the diameter.

Practice

5. a. 37.68 cm
  b. 113.04 cm^2
7. a. 13.71 in.
  b. 12.5325 in.^{2}

Applications

9. a. 63.6 square inches
  b. $0.13 per square inch

Writing & Thinking

11. One diameter is equal to two radii. Thus, $d = 2r$ and $C = 2\pi r = \pi d$.

6.5 Exercises

Concept Check

1. True
3. True
5. a. C
  b. E
  c. D
  d. B
  e. A

Practice

7. 12.56 mm^3
9. 1017.36 mm^3

Applications

11. 800 ft^2

Writing & Thinking

13. Volume is measured in cubic units. Volume takes up a three-dimensional space and the units can be thought of as small cubes which leads to the concept of cubic units.

6.6 Exercises

Concept Check

1. False; Similar triangles have corresponding sides that are proportional.
3. False; If $\triangle ABC = \triangle DEF$, then $AC = DF$.

Practice

5. The triangles are not similar. The corresponding sides are not proportional.
7. $x = 50°$; $y = 70°$
9. Congruent by SAS

Applications

11. 7.5 feet

Writing & Thinking

13. a. ∠B should be equal to m∠E and m∠C should be equal to m∠F.
  b. Corresponding sides are proportional, not that they have the same length.

6.7 Exercises

Concept Check

1. True
3. True

Practice

5. 6
7. Yes, $64 + 36 = 100$
9. $c = 5$

Applications

11. 17.0 inches

Chapter 7: Statistics, Graphs, and Probability

7.1 Exercises

Concept Check

1. True
3. False; The number that appears the greatest number of times in a set of data is the mode.

Practice

5. a. 58
  b. 57
  c. 57
  d. 4
7. a. $48,625
  b. $46,500
  c. $63,000
  d. $43,000

Applications

9. 79

Writing & Thinking

11. The mean and median may be the same number as in the set of data: 12, 16, 20. However, more generally, the mean and median are two different numbers as in the set of data: 22, 23, 36. In this set, the median is 23, while the mean is 27.

7.2 Exercises

Concept Check

1. True
3. True

Applications

5. a. Social Science
   b. Chemistry & Physics
   c. About 3300
   d. About 21.2%

7. a. February and May
   b. 6 inches
   c. March
   d. 3.58 inches

Writing & Thinking

9. All graphs should be 1. clearly labeled, 2. easy to read, and 3. have appropriate titles.

7.3 Exercises

Concept Check

1. False; In creating a vertical bar graph, all bar widths should be the same.
3. True

Applications

5.

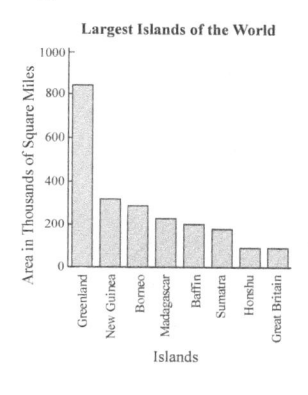

Largest Islands of the World

7.

Percent of Population with Particular Blood Types

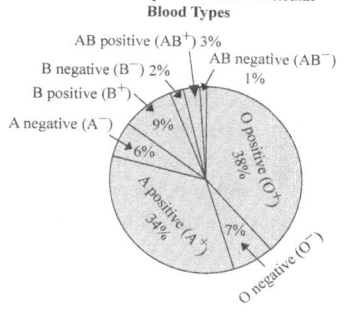

Writing & Thinking

9. Constructing a graph would require a thorough understanding of the data and concepts represented in the graph as well as the proper type of graph to best communicate the information.

7.4 Exercises

Concept Check

1. False; The individual result of an experiment is an outcome.
3. True

Applications

5.

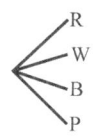

$S = \{R, W, B, P\}$
R = red, W = white,
B = blue, P = purple

7. $\dfrac{2}{5}$

Writing & Thinking

9. Chance experiments include, but are not limited to, tossing a coin, spinning a bottle, drawing a card from a standard deck of cards, picking numbers in the lottery, choosing straws, and picking colored marbles.

Chapter 8: Introduction to Algebra

8.1 Exercises

Concept Check

1. True
3. True
5. [number line with points at −3, −2, 0, 1]
7. 0, 4, 8
9. True

Applications

11. −4500 meters

Writing & Thinking

13. If y is a negative number then $-y$ represents a positive number. For example, if $y = -2$, then $-y = -(-2) = 2$.

8.2 Exercises

Concept Check

1. False; The sum of a positive and negative number can be positive, negative, or zero.
3. False; The sum of two positive numbers is always positive, zero is neither positive nor negative.

Practice

5. −6
7. −0.5

Applications

9. a. $45,000 + (−$8000) + (−$2000) + $15,000
   b. $50,000

Writing & Thinking

11. $|0| + |0| = 0$

8.3 Exercises

Concept Check

1. False; The sum of a number and its additive inverse is zero.
3. True

Practice

5. −11
7. 3
9. −15
11. −18°F (a decrease of 18 degrees Fahrenheit)

Writing & Thinking

13. Add the opposite of the second number to the first number.

8.4 Exercises

Concept Check

1. True

9.2 Exercises

3. False; The product and quotient will be positive.

Practice
5. 48
7. 2

Applications
9. −12

Writing & Thinking
11. Negative; The product of every two negative numbers will be positive and this result multiplied by the remaining negative will give a negative answer.

8.5 Exercises

Concept Check
1. True
3. True

Practice
5. a. 36
  b. 16
7. −10
9. 129

Applications
11. a. −$42 − $35 − (3 · $5)
  b. −$92

Writing & Thinking
13. Smaller; When any positive number is multiplied by a fraction (or decimal) between 0 and 1, the result will be smaller. This is what is happening when a number between 0 and 1 is squared. Answers will vary.

8.6 Exercises

Concept Check
1. False; The commutative property of addition allows the order to change.
3. False; The additive identity of all numbers is 0.

Practice
5. 3 + 7
7. 4 · 19
9. (16 + 9) + 11

Applications
11. a. $118.25
  b. $11 · $\left(6\frac{1}{2}\right)$ + $11 · $\left(4\frac{1}{4}\right)$
  c. Distributive property.

Writing & Thinking
13. a. Multiply each term that is part of the sum (in the parentheses) by a.
  b. An expression that "distributes addition over multiplication" would look like $a + (bc) = (a+b)(a+c)$, however this is not a true statement. Answers will vary.

8.7 Exercises

Concept Check
1. True
3. False; In the term "12a," 12 is the coefficient.

Practice
5. −5, 3, and 8 are like terms; 7x and 9x are like terms.
7. 10x
9. 3x + 4; 13

Applications
11. $50,000

Writing & Thinking
13. Like terms have the same variables with the same exponents. For example, $4a^2bc^3$ and $−3a^2bc^3$ are like terms. Unlike terms either have different variables or possibly the same variables with different exponents. For example 6ab and −9a^2b are unlike terms and 5xy and 13ax are unlike terms.

8.8 Exercises

Concept Check
1. True
3. False; Subtraction is indicated by the phrase "five less than a number."

Practice
5. x + 6
7. $\frac{x}{2}$ − 18
9. a. 4n − 6
  b. 6 − 4n
11. The product of a number and negative nine

Writing & Thinking
13. The Commutative Property of Addition and Multiplication permits the order of items being added or multiplied to change and still have the same result. This property does not hold true for subtraction or division. Therefore, order is important for subtraction and division problems or the answer will change or be incorrect.

Chapter 9: Solving Linear Equations and Inequalities

9.1 Exercises

Concept Check
1. True
3. True

Practice
5. x = 1 is not a solution
7. x = 7
9. y = 3

Applications
11. 1945 *kanji* characters

Writing & Thinking
13. a. Yes. It is stating that 6 + 3 is equal to 9.
  b. No. If we substitute 4 for x, we get the statement 9 = 10, which is not true.

9.2 Exercises

Concept Check
1. False; The addition and multiplication principles of equality can be used with decimal or fractional coefficients.
3. True

Practice
5. x = −3
7. $x = -\frac{27}{10}$

Applications
9. 14,000 tickets per hour

Writing & Thinking
11. a. The 4 should have been multiplied by 3 so that the 3 was distributed over the entire left-hand side of the equation; Correct answer is x = 15.
  b. 3 should be subtracted from each side, not from each term, and 5x − 3 doesn't simplify to 2x; Correct answer is $x = \frac{8}{5}$.

9.3 Exercises

Concept Check

1. True
3. False; It is called a contradiction.

Practice

5. $x = -5$
7. $x = -1$
9. Contradiction

Applications

11. 20 guests

Writing & Thinking

13. a. $5x + 1$
    b. $x = 6$
    c. Answers will vary.

9.4 Exercises

Concept Check

1. False; Case matters in formulas.
3. True

Applications

5. $1030
7. $b = P - a - c$
9. $t = \dfrac{I}{Pr}$

9.5 Exercises

Concept Check

1. True
3. False; Odd integers are integers that are not even.

Practice

5. $x - 5 = 13 - x$; 9

7. $n + (n+1) + (n+2) = 93$; 30, 31, 32

Applications

9. a. The unknown value is the number of postcards purchased.
   b. $p = 9$
   c. Brooke purchased 9 postcards.

Writing & Thinking

11. a. $n, n+2, n+4, n+6$
    b. $n, n+2, n+4, n+6$
    c. Yes. Answers will vary.

9.6 Exercises

Concept Check

1. False; The value of r should be written as a decimal number.
3. True

Applications

5. 8.75 hours
7. $112.50

Writing & Thinking

9. If each equal side is 9 cm long, that would make the perimeter more than 18 cm; Correct answer: 6 cm

9.7 Exercises

Concept Check

1. True
3. False; Only one value in the solution set needs to be checked.

Practice

5. Half-open interval
7. $(4, \infty)$
9. $[4, \infty)$

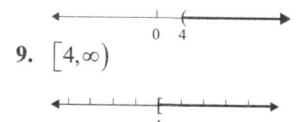

Applications

11. a. The student would need a score higher than 102 points, which is not possible. Thus he cannot earn an A in the course.
    b. The student must score at least 192 points to earn an A in the course.

Writing & Thinking

13. a. Answers will vary.
    b. Answers will vary.

9.8 Exercises

Concept Check

1. False; The union of two sets contains elements that belong to either one set, the other set, or both sets.
3. True

Practice

5. Union: {1, 2, 3, 4, 6, 8}; Intersection: {2, 4}
7. $(-1, 6)$
9. $(-9, 1)$

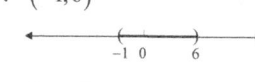

9.9 Exercises

Concept Check

1. False; Equations involving absolute value can have more than one solution.
3. True

Practice

5. No solution
7. No solution
9. $x = -10, \dfrac{18}{7}$

9.10 Exercises

Concept Check

1. False; Only one statement/inequality must be true.
3. False; Must be greater than 2

Practice

5. $(-\infty, \infty)$
7. No solution
9. $\left(-\infty, -\dfrac{11}{3}\right] \cup [3, \infty)$

Writing & Thinking

11. a.
    b. $|x| \leq 10$
    c. $[-10, 10]$, Closed interval

Chapter 10: Graphing Linear Equations and Inequalities

10.1 Exercises

Concept Check

1. True
3. True

Practice

5. $\begin{cases} A(-5, 1), B(-3, 3), \\ C(-1, 1), D(1, 2), \\ E(2, -2) \end{cases}$

7. a. $(0, -1)$
   b. $(4, 1)$
   c. $(2, 0)$
   d. $(8, 3)$

9. b, c

Application

11. a.
D	E
100	85
200	170
300	255
400	340
500	425

b. [graph]

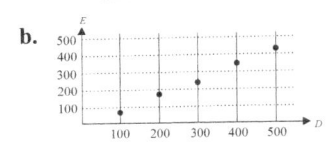

10.2 Exercises

Concept Check

1. True
3. False; Horizontal lines have y-intercepts

Practice

5. [graph: $x + y = 3$, (0, 3), (3, 0)]
7. [graph: $y = -3$, (-5, -3), (0, -3)]
9. [graph: $3x - 7y = -21$, (-7, 0), (0, 3)]

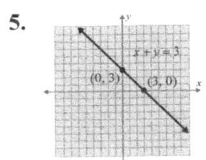

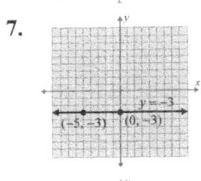

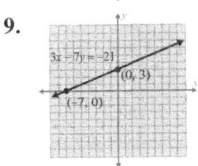

Application

11. The y-intercept is $(0, 30)$, meaning that if a student does no homework at all, the student will get a score of 30 points on the exam.

Writing & Thinking

13. Substitute the x and y values into the equation. Then evaluate both sides to see if the equation is true.

10.3 Exercises

Concept Check

1. True
3. True

Practice

5. m is undefined
7. [graph] Vertical line; m is undefined
9. [graph: y-int = (0, 3), $y = 3$, $m = 0$]

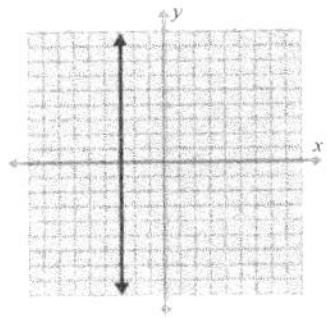

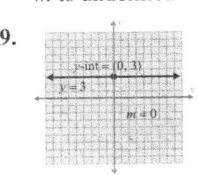

Applications

11. $4000/year [graph: Car Value vs Year]

Writing & Thinking

13. a. For any horizontal line, all of the y values will be the same. Thus the formula for slope will always have 0 in the numerator making the slope of every horizontal line 0.

b. For any vertical line, all of the x values will be the same. Thus the formula for slope will always have 0 in the denominator making the slope of every vertical line undefined.

10.4 Exercises

Concept Check

1. True
3. True

Practice

5. a. $m = 2$ b. $(3, 1)$
c. [graph]
7. $y = -\dfrac{1}{7}x + \dfrac{2}{7}$
9. $y = 6$

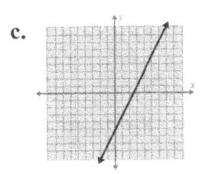

Application

11. a. $P = 100t - 5000$
b. 50 tickets

10.5 Exercises

Concept Check

1. True
3. True

Practice

5. $\{(-5, -4), (-4, -2), (-2, -2), (1, -2), (2, 1)\}$;
$D = \{-5, -4, -2, 1, 2\}$;
$R = \{-4, -2, 1\}$;
Function

7. Not a function;
$D = (-\infty, \infty)$;
$R = (-\infty, \infty)$

9. a. -10 b. 86 c. 86

10.6 Exercises

Concept Check

1. True
3. False; The boundary line is solid if the inequality uses $\leq$ or $\geq$.

Practice

5. [graph]
7. [graph]
9. [graph]

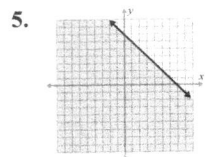

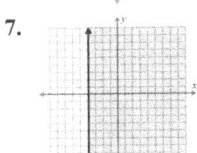

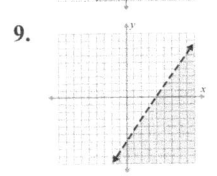

Writing & Thinking

11. Test any point not on the line. If the test point satisfies the inequality, shade the half-plane on that side of the line. Otherwise, shade the other half-plane.

Geometry

P = Perimeter, A = Area, C = Circumference, V = Volume, SA = Surface Area

Perimeter and Area

Rectangle	Square	Triangle	Parallelogram	Trapezoid	Circle
$P = 2l + 2w$	$P = 4s$	$P = a + b + c$	$P = 2a + 2b$	$P = a + b + c + d$	$C = 2\pi r = \pi d$
$A = lw$	$A = s^2$	$A = \frac{1}{2}bh$	$A = bh$	$A = \frac{1}{2}h(b+c)$	$A = \pi r^2$

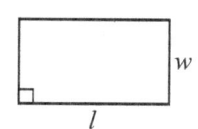

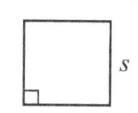

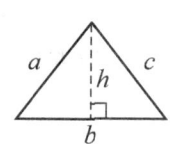

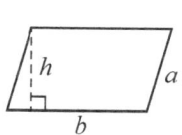

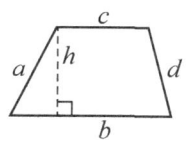

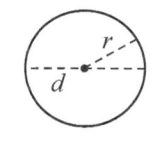

Volume and Surface Area

Rectangular Solid	Rectangular Pyramid	Right Circular Cone	Right Circular Cylinder	Sphere
$V = lwh$	$V = \frac{1}{3}lwh$	$V = \frac{1}{3}\pi r^2 h$	$V = \pi r^2 h$	$V = \frac{4}{3}\pi r^3$
$SA = 2lw + 2wh + 2lh$			$SA = 2\pi r^2 + 2\pi rh$	$SA = 4\pi r^2$

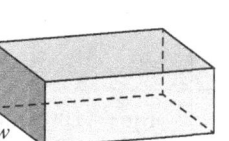

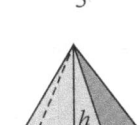

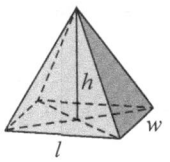

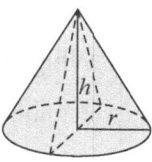

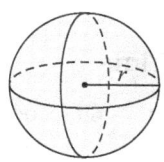

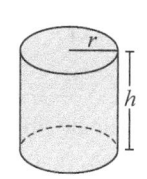

Angles Classified by Measure

Acute
$0° < m\angle A < 90°$

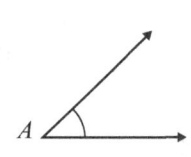

Right
$m\angle A = 90°$

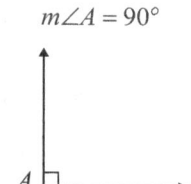

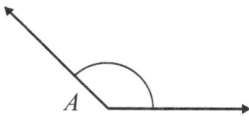

Obtuse
$90° < m\angle A < 180°$

Straight
$m\angle A = 180°$

Triangles Classified by Sides

Scalene
No two sides have equal lengths.

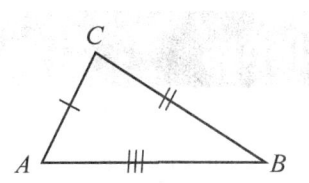

Isosceles
At least two sides have equal lengths.

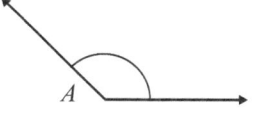

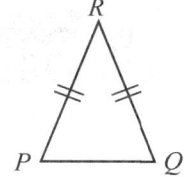

Equilateral
All three sides have equal lengths.

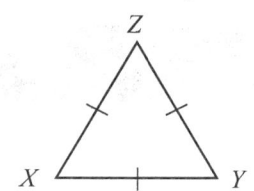

Triangles Classified by Angles

Acute
All three angles are acute.

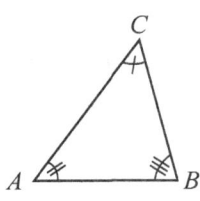

Right
One angle is a right angle.

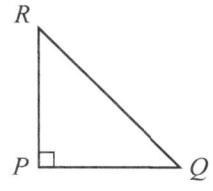

Obtuse
One angle is obtuse.

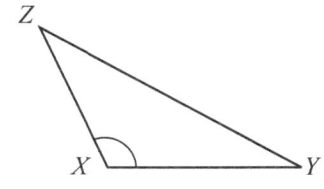

US Customary System of Measurement

Length

12 inches (in.) = 1 foot (ft)
36 inches = 1 yard (yd)
3 feet = 1 yard
5280 feet = 1 mile (mi)

Capacity

8 fluid ounces (fl oz) = 1 cup (c)
2 cups = 1 pint (pt) = 16 fluid ounces
2 pints = 1 quart (qt)
4 quarts = 1 gallon (gal)

Weight

16 ounces (oz) = 1 pound (lb)
2000 pounds = 1 ton (T)

Time

60 seconds (sec) = 1 minute (min)
60 minutes = 1 hour (hr)
24 hours = 1 day
7 days = 1 week

Temperature

Celsius (C) to Fahrenheit (F)

$$F = \frac{9C}{5} + 32$$

Fahrenheit (F) to Celsius (C)

$$C = \frac{5(F - 32)}{9}$$

Metric System of Measurement

Length

1 millimeter (mm)	= 0.001 meter	1 m = 1000 mm
1 centimeter (cm)	= 0.01 meter	1 m = 100 cm
1 decimeter (dm)	= 0.1 meter	1 m = 10 dm
1 meter (m)	= 1.0 meter	
1 dekameter (dam)	= 10 meters	1 cm = 10 mm
1 hectometer (hm)	= 100 meters	
1 kilometer (km)	= 1000 meters	

Capacity (Liquid Volume)

1 milliliter (mL)	= 0.001 liter	1 L = 1000 mL
1 liter (L)	= 1.0 liter	
1 hectoliter (hL)	= 100 liters	
1 kiloliter (kL)	= 1000 liters	1 kL = 10 hL

Weight

1 milligram (mg)	= 0.001 gram	1 g = 1000 mg
1 centigram (cg)	= 0.01 gram	
1 decigram (dg)	= 0.1 gram	
1 gram (g)	= 1.0 gram	
1 dekagram (dag)	= 10 grams	
1 hectogram (hg)	= 100 grams	
1 kilogram (kg)	= 1000 grams	1 g = 0.001 kg
1 metric ton (t)	= 1000 kilograms	1 kg = 0.001 t

1 t = 1000 kg = 1,000,000 g = 1,000,000,000 mg

Land Area

1 are (a) = 100 m^2
1 hectare (ha) = 100 a = 10 000 m^2

US Customary and Metric Equivalents

Length

1 in. = 2.54 cm (exact)	1 cm ≈ 0.394 in.
1 ft ≈ 0.305 m	1 m ≈ 3.28 ft
1 yd ≈ 0.914 m	1 m ≈ 1.09 yd
1 mi ≈ 1.61 km	1 km ≈ 0.62 mi

Area

1 in.^{2} ≈ 6.45 cm^2	1 cm^2 ≈ 0.155 in.^{2}
1 ft^2 ≈ 0.093 m^2	1 m^2 ≈ 10.764 ft^2
1 yd^2 ≈ 0.836 m^2	1 m^2 ≈ 1.196 yd^2
1 acre ≈ 0.405 ha	1 ha ≈ 2.47 acres

Volume

1 qt ≈ 0.946 L	1 L ≈ 1.06 qt
1 gal ≈ 3.785 L	1 L ≈ 0.264 gal

Mass

1 oz ≈ 28.35 g	1 g ≈ 0.035 oz
1 lb ≈ 0.454 kg	1 kg ≈ 2.205 lb

Notation and Terminology

Exponents

$$\underbrace{a \cdot a \cdot a \cdot a \cdot \ldots \cdot a}_{n \text{ factors}} = a^n$$

where n is the exponent and a is the base.

Fractions

$$\frac{a}{b} \begin{array}{l} \leftarrow \text{numerator} \\ \leftarrow \text{denominator} \end{array}$$

Least Common Multiple (LCM)

Given a set of counting numbers, the smallest number that is a multiple of each of these numbers.

Ratios

$$\frac{a}{b} \quad \text{or} \quad a:b \quad \text{or} \quad a \text{ to } b$$

A comparison of two quantities by division.

Proportions

$$\frac{a}{b} = \frac{c}{d} \quad \text{A statement that two ratios are equal.}$$

Greatest Common Factor (GCF)

Given a set of integers, the largest integer that is a factor (or divisor) of all of the integers.

Types of Numbers

Natural Numbers (Counting Numbers):
$\mathbb{N} = \{1, 2, 3, 4, 5, 6, \ldots\}$

Whole Numbers: $\mathbb{W} = \{0, 1, 2, 3, 4, 5, 6, \ldots\}$

Integers: $\mathbb{Z} = \{\ldots, -4, -3, -2, -1, 0, 1, 2, 3, 4, \ldots\}$

Rational Numbers: A number that can be written in the form $\frac{a}{b}$ where a and b are integers and $b \neq 0$.

Irrational Numbers: A number that can be written as an infinite nonrepeating decimal.

Real Numbers: All rational and irrational numbers.

Complex Numbers: All real numbers and the even roots of negative numbers. The standard form of a complex number is $a + bi$, where a and b are real numbers, a is called the real part and b is called the imaginary part.

Absolute Value

$|a|$ The distance a real number a is from 0.

Equality and Inequality Symbols

$=$	"is equal to"
$\neq$	"is not equal to"
$<$	"is less than"
$>$	"is greater than"
$\leq$	"is less than or equal to"
$\geq$	"is greater than or equal to"

Sets

The **empty set** or **null set** (symbolized $\emptyset$ or $\{\ \}$): A set with no elements.

The **union** of two (or more) sets (symbolized $\cup$): The set of all elements that belong to either one set or the other set or to both sets.

The **intersection** of two (or more) sets (symbolized $\cap$): The set of all elements that belong to both sets.

The word **or** is used to indicate union and the word **and** is used to indicate intersection.

Algebraic and Interval Notation for Intervals

Type of Interval	Algebraic Notation	Interval Notation	Graph
Open Interval	$a < x < b$	(a, b)	
Closed Interval	$a \leq x \leq b$	$[a, b]$	
Half-Open Interval	$\begin{cases} a \leq x < b \\ a < x \leq b \end{cases}$	$[a, b)$ $(a, b]$	
Open Interval	$\begin{cases} x > a \\ x < b \end{cases}$	(a, ∞) $(-\infty, b)$	
Half-Open Interval	$\begin{cases} x \geq a \\ x \leq b \end{cases}$	$[a, \infty)$ $(-\infty, b]$	

Radicals

The symbol $\sqrt{}$ is called a **radical sign**.
The number under the radical sign is called the **radicand**.

The complete expression, such as $\sqrt{64}$, is called a **radical** or **radical expression**.

In a cube root expression $\sqrt[3]{a}$, the number 3 is called the index. In a square root expression such as $\sqrt{a}$, the index is understood to be 2 and is not written.

The Imaginary Number i

$$i = \sqrt{-1} \quad \text{and} \quad i^2 = \left(\sqrt{-1}\right)^2 = -1$$

Formulas and Theorems

Percent

$\dfrac{P}{100} = \dfrac{A}{B}$ (the percent proportion),

where

$P\%$ = percent (written as the ratio $\dfrac{P}{100}$)
B = base (number we are finding the percent of)
A = amount (a part of the base)

$R \cdot B = A$ (the basic percent equation),

where

R = rate or percent (as a decimal or fraction)
B = base (number we are finding the percent of)
A = amount (a part of the base)

Profit

Profit: The difference between selling price and cost.

$$\text{Profit} = \text{Selling Price} - \text{Cost}$$

Percent of Profit:

1. Percent of profit **based on cost**: $\dfrac{\text{Profit}}{\text{Cost}}$

2. Percent of profit **based on selling price**: $\dfrac{\text{Profit}}{\text{Selling Price}}$

Interest

Simple Interest: $I = P \cdot r \cdot t$

Compound Interest: $A = P\left(1 + \dfrac{r}{n}\right)^{nt}$

Continuously Compounded Interest: $A = Pe^{rt}$

where

I = interest (earned or paid)
A = amount accumulated
P = principal (the amount invested or borrowed)
r = annual interest rate in decimal or fraction form
t = time (years or fraction of a year)
n = number of compounding periods in 1 year
$e = 2.718281828459\ldots$

The Pythagorean Theorem

In a right triangle, the square of the length of the hypotenuse is equal to the sum of the squares of the lengths of the two legs. $c^2 = a^2 + b^2$

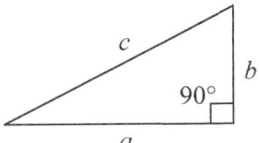

Probability of an Event

$$\text{probability of an event} = \dfrac{\text{number of outcomes in event}}{\text{number of outcomes in sample space}}$$

Distance-Rate-Time

$d = rt$ The **distance traveled** d equals the product of the rate of speed r and the time t.

Special Products

1. $x^2 - a^2 = (x+a)(x-a)$ Difference of two squares
2. $x^2 + 2ax + a^2 = (x+a)^2$ Square of a binomial sum
3. $x^2 - 2ax + a^2 = (x-a)^2$ Square of a binomial difference
4. $x^3 + a^3 = (x+a)(x^2 - ax + a^2)$ Sum of two cubes
5. $x^3 - a^3 = (x-a)(x^2 + ax + a^2)$ Difference of two cubes

Change-of-Base Formula for Logarithms

For $a, b, x > 0$ and $a, b \neq 1$, $\log_b x = \dfrac{\log_a x}{\log_a b}$.

Distance Between Two Points

The distance d between points $P(x_1, y_1)$ and $Q(x_2, y_2)$ is $d = \sqrt{(x_2 - x_1)^2 + (y_2 - y_1)^2}$.

Midpoint Formula

The midpoint between points $P(x_1, y_1)$ and $Q(x_2, y_2)$ is $\left(\dfrac{x_1 + x_2}{2}, \dfrac{y_1 + y_2}{2}\right)$.

Principles and Properties

Properties of Addition and Multiplication

Property	Addition	Multiplication
Commutative Property	$a + b = b + a$	$ab = ba$
Associative Property	$(a+b)+c = a+(b+c)$	$a(bc) = (ab)c$
Identity	$a + 0 = 0 + a = a$	$a \cdot 1 = 1 \cdot a = a$
Inverse	$a + (-a) = 0$	$a \cdot \dfrac{1}{a} = 1 \; (a \neq 0)$

Zero-Factor Law: $a \cdot 0 = 0 \cdot a = 0$

Distributive Property: $a(b+c) = a \cdot b + a \cdot c$

Addition (or Subtraction) Principle of Equality

$A = B$, $A + C = B + C$, and $A - C = B - C$ have the same solutions (where A, B, and C are algebraic expressions).

Multiplication (or Division) Principle of Equality

$A = B$, $AC = BC$, and $\dfrac{A}{C} = \dfrac{B}{C}$ have the same solutions (where A and B are algebraic expressions and C is any nonzero constant, $C \neq 0$).

Properties of Exponents

For nonzero real numbers a and b and integers m and n:

The exponent 1	$a = a^1$
The exponent 0	$a^0 = 1$
The product rule	$a^m \cdot a^n = a^{m+n}$
The quotient rule	$\dfrac{a^m}{a^n} = a^{m-n}$
Negative exponents	$a^{-n} = \dfrac{1}{a^n}$
Power rule	$(a^m)^n = a^{mn}$
Power of a product	$(ab)^n = a^n b^n$
Power of a quotient	$\left(\dfrac{a}{b}\right)^n = \dfrac{a^n}{b^n}$

Zero-Factor Law

If a and b are real numbers, and $a \cdot b = 0$, then $a = 0$ or $b = 0$ or both.

Properties of Rational Numbers (or Fractions)

If $\dfrac{P}{Q}$ is a rational expression and P, Q, R, and K are polynomials where $Q, R, S, K \neq 0$, then

The Fundamental Principle	$\dfrac{P}{Q} = \dfrac{P \cdot K}{Q \cdot K}$
Multiplication	$\dfrac{P}{Q} \cdot \dfrac{R}{S} = \dfrac{P \cdot R}{Q \cdot S}$
Division	$\dfrac{P}{Q} \div \dfrac{R}{S} = \dfrac{P}{Q} \cdot \dfrac{S}{R}$
Addition	$\dfrac{P}{Q} + \dfrac{R}{Q} = \dfrac{P+R}{Q}$
Subtraction	$\dfrac{P}{Q} - \dfrac{R}{Q} = \dfrac{P-R}{Q}$

Properties of Radicals

If a and b are positive real numbers, n is a positive integer, m is any integer, and $\sqrt[n]{a}$ is a real number then

1. $\sqrt[n]{ab} = \sqrt[n]{a} \cdot \sqrt[n]{b}$
2. $\sqrt[n]{\dfrac{a}{b}} = \dfrac{\sqrt[n]{a}}{\sqrt[n]{b}}$
3. $\sqrt[n]{a} = a^{\frac{1}{n}}$
4. $a^{\frac{m}{n}} = \left(a^{\frac{1}{n}}\right)^m = (a^m)^{\frac{1}{n}}$

or, in radical notation,

$a^{\frac{m}{n}} = \left(\sqrt[n]{a}\right)^m = \sqrt[n]{a^m}$

Properties of Logarithms

For $b > 0$, $b \neq 1$, $x, y > 0$, and any real number r,

1. $\log_b 1 = 0$
2. $\log_b b = 1$
3. $x = b^{\log_b x}$
4. $\log_b b^x = x$
5. $\log_b xy = \log_b x + \log_b y$ The product rule
6. $\log_b \dfrac{x}{y} = \log_b x - \log_b y$ The quotient rule
7. $\log_b x^r = r \cdot \log_b x$ The power rule

Properties of Equations with Exponents and Logarithms

For $b > 0$, $b \neq 1$,

1. If $b^x = b^y$, then $x = y$.
2. If $x = y$, then $b^x = b^y$.
3. If $\log_b x = \log_b y$, then $x = y$ ($x > 0$ and $y > 0$).
4. If $x = y$, then $\log_b x = \log_b y$ ($x > 0$ and $y > 0$).

Notes

Notes

Notes

Notes

Notes